GeoPlanet: Earth and Planetary Sciences

More information about this series at http://www.springer.com/series/8821

Barbara Guterch · Jan Kozák
Editors

Studies of Historical Earthquakes in Southern Poland

Outer Western Carpathian Earthquake of December 3, 1786, and First Macroseismic Maps in 1858–1901

Springer

Editors
Barbara Guterch
Institute of Geophysics
Polish Academy of Sciences
Warsaw
Poland

Jan Kozák
Institute of Geophysics
Academy of Sciences of the Czech Republic
Prague
Czech Republic

The GeoPlanet: Earth and Planetary Sciences Book Series is in part a continuation of Monographic Volumes of Publications of the Institute of Geophysics, Polish Academy of Sciences, the journal published since 1962 (http://pub.igf.edu.pl/index.php).

ISSN 2190-5193 ISSN 2190-5207 (electronic)
GeoPlanet: Earth and Planetary Sciences
ISBN 978-3-319-15445-9 ISBN 978-3-319-15446-6 (eBook)
DOI 10.1007/978-3-319-15446-6

Library of Congress Control Number: 2015932247

Springer Cham Heidelberg New York Dordrecht London

Printed on acid-free paper

Springer International Publishing AG Switzerland is part of Springer Science+Business Media (www.springer.com)

Series Editors

Managing Editor

Anna Dziembowska
Institute of Geophysics, Polish Academy of Sciences

Advisory Board

Kon-Kee Liu
Institute of Hydrological
and Oceanic Sciences
National Central University Jhongli
Jhongli, Taiwan

Teresa Madeyska
Research Centre in Warsaw
Institute of Geological Sciences
Warszawa, Poland

Stanisław Massel
Institute of Oceanology
Polish Academy of Sciences
Sopot, Poland

Antonio Meloni
Instituto Nazionale di Geofisica
Rome, Italy

Evangelos Papathanassiou
Hellenic Centre for Marine Research
Anavissos, Greece

Kaja Pietsch
AGH University of Science and
Technology
Kraków, Poland

Dušan Plašienka
Prírodovedecká fakulta, UK
Univerzita Komenského
Bratislava, Slovakia

Barbara Popielawska
Space Research Centre
Polish Academy of Sciences
Warszawa, Poland

Tilman Spohn
Deutsches Zentrum für Luftund
Raumfahrt in der Helmholtz
Gemeinschaft
Institut für Planetenforschung
Berlin, Germany

Krzysztof Stasiewicz
Swedish Institute of Space Physics
Uppsala, Sweden

Ewa Szuszkiewicz
Department of Astronomy
and Astrophysics
University of Szczecin
Szczecin, Poland

Roman Teisseyre
Department of Theoretical Geophysics
Institute of Geophysics
Polish Academy of Sciences
Warszawa, Poland

Jacek Tronczynski
Laboratory of Biogeochemistry
of Organic Contaminants
IFREMER DCN_BE
Nantes, France

Steve Wallis
School of the Built Environment
Heriot-Watt University
Riccarton, Edinburgh
Scotland, UK

Wacław M. Zuberek
Department of Applied Geology
University of Silesia
Sosnowiec, Poland

Piotr Życki
Nicolaus Copernicus Astronomical
Centre
Polish Academy of Sciences
Warszawa, Poland

Foreword

The interest in seismicity of Poland, generally considered as seismically quiet, has recently grown on account of the two unexpected events that happened in this area: the earthquake of Kaliningrad (about 60 km NE of the Polish border) in 2004 and the one near Jarocin (central Poland) in 2012. This indicates how important it is to revisit the seismic history in this part of Europe, and currently update our knowledge about it, not only for purely scientific but also quite practical reasons.

The articles presented in this book partly fill up the gaps in our knowledge by providing some new historical data on the seismicity in Poland. The book is mainly focused on southern Poland, and presents the facts about the earthquakes felt there, notably the most spectacular one of December 3, 1786, in the Outer Western Carpathians, which affected the historical town of Kraków. It had been fairly well documented in the past, but many new data were found as a result of many-year search over the archives in Kraków and elsewhere. We admire the hard work of the authors, Barbara Guterch, Hanna Lewandowska-Marciniak and Maciej Mazur, in going through these documents, often not registered adequately in the archives or located in quite unexpected places. The manuscripts and articles/notes in contemporary journals are quoted in original languages (Polish, Latin, German, French).

The historical part of the book provides also a unique collection of 19 macroseismic intensity maps, compiled in the second half of the nineteenth century, forerunners of the contemporary maps of seismicity. These valuable materials were collected and analysed by Prof. Jan Kozak from the Geophysical Institute of the Czech Academy of Sciences, world acknowledged expert in seismology and an ardent collector of old prints and maps.

Two out of five chapters give information updated to the present times. These are the seismological catalog of Poland until the year 2014, an amended version of the catalog published in 2009 in Polish, and the article entitled "Seismicity of Polish part of the Western Carpathians in the light of recent data" which also gives an account of recent seismic monitoring of the country carried out by the Institute of Geophysics, Polish Academy of Sciences.

The full documentation of Polish instrumental seismological data until the year 2006 is available in printed form, first as bulletins of individual observatories, and then in the framework of the series Publications of the Institute of Geophysics, Polish Academy of Sciences. In 2007 we stopped issuing printed versions, leaving the Internet access to the data. The publication of bulletins was for many years supervised by the late Prof. S.J. Gibowicz, former Head of the Seismological Department of the Institute, who also drew the attention of seismologists to the unusual event of the December 3, 1786. He was mainly engaged in the study of mining-induced seismicity, but this is out of the scope of the present book.

Although mostly history-oriented, the book is very timely and may be of some help right now; it provides important information to be considered in seismic hazard assessment for planning new industrial plants, dwelling houses and other edifices.

Institute of Geophysics
Polish Academy of Sciences

Anna Dziembowska
Roman Teisseyre

Preface

This monograph presents early studies and documents on historical earthquakes in Central Europe, in particular in the border regions between the three countries: Poland, Czech Republic and Slovakia. These regions, embracing the Western Carpathians and the Sudetes, are a site of considerable geotectonic activity. The monograph is dedicated to the recent seismology; it complies with the recommendations of the European Seismological Commission of 1986 expressing the need for historical earthquake research. In such research, the macroseismic maps for the second half of the nineteenth century and archival documents for the late eighteenth century, presented in the monograph, have the same value as the seismograms in the instrumental era and provide data for seismic hazard assessment. The knowledge of seismic hazard is of utmost importance for the contemporary huge industrial projects in Europe; its detailed recognition is a must, notably in the areas where nuclear power plants, water power plants and other investments are planned, for which the seismic hazard should be determined as well as possible.

The documentation on seismic activity presented in the monograph provides a background for systematic monitoring of selected regions. We should keep in mind that there are no fully aseismic regions, as still evidenced by new events and newly retrieved historical data. Earthquakes of 2004 in the vicinity of Kaliningrad, 60 km NE from the Polish border, are a good example of events that unexpectedly occurred in the area regarded as quite aseismic.

Barbara Guterch

Contents

Chapter 1
The Earthquake in the Outer Western Carpathians, Poland, on December 3, 1786

Barbara Guterch, Hanna Lewandowska-Marciniak and Maciej Mazur

Abstract The aim of this study is a reinterpretation of the December 3, 1786, earthquake in the Outer Western Carpathians using known and newly found documents. The earthquake completed a set of events in the exceptionally active two-year period 1785–1786, when magnitudes and depths of epicenters reached, respectively, $M_W = 5.3$ and $h = 35$ km, being the strongest and deepest seismic events ever recorded in this region. Prior to the second half of the XVIII century, no earthquakes have been recognized; instrumentally, only 2 events, of $M_W = 2.4$ and $M_W = 2.6$, were recognized in the year 1994. Special searches were carried out in ecclesial and municipal archives to find unknown documents concerning the earthquake of December 3, 1786. New macroseismic maps have been compiled from qualitative data assessment of varying degrees of source significance and seismological importance. The mean parameters of the earthquake have been determined. High dispersion of intensity data points let us assume that it might have been not a single earthquake but a seismic swarm, with epicenters supposedly in the Upper Silesia, the Outer Western Carpathians, and the Inner Western Carpathians. The main earthquake in the years 1785–1786 occurred in the area of the Upper Silesian Block close to the tectonic junction of the Carpathian and Variscan orogenic fronts. References of the earthquake in manuscripts and contemporary journals, in the original languages as well as a catalogue of all localities where the quake was felt, including quotations translated into English, have been presented in a separate chapter at the end of this book.

Keywords Quality of macroseismic records · Intensity data points · Macroseismic paramerters of earthquake

---

B. Guterch (✉) · H. Lewandowska-Marciniak · M. Mazur
Institute of Geophysics, Polish Academy of Sciences,
ul. Księcia Janusza 64, 01-452 Warsaw, Poland
e-mail: bguterch@igf.edu.pl

B. Guterch and J. Kozák (eds.), *Studies of Historical Earthquakes in Southern Poland*, GeoPlanet: Earth and Planetary Sciences,
DOI 10.1007/978-3-319-15446-6_1

1.1 Introduction

Three main earthquakes occurred in the Outer Western Carpathians in the years 1785–1786, i.e., on August 22, 1785; February 27, 1786; and December 3, 1786; and they have remained to be the strongest, sufficiently well documented, seismic events ever recorded in the region. The earthquakes of years 1785–1786 caused a lot of concern to the people not accustomed to such events. Especially the last one, on December 3, 1786, that seemed to be the strongest, captured people's attention.

After the Great Lisbon earthquake of November 1, 1755, strongly felt in Central Europe, people's consciousness about earthquake effects even in such "aseismic" countries like Poland, has increased. Newspapers in Warsaw (Gazeta Warszawska 1786), Wrocław (Schlesische Privilegirte Zeitung 1786), Oleśnica (Schlesische Provinzialblätter 1786), Wilno (Gazeta Wileńska 1786), Lwów (Lwowskie Tygodniowe Wiadomości 1786, 1787), Brno (Brünner Zeitung 1786) reported the event. Even prestigious Nova Acta Academiae Scientiarum Imperialis Petropolitanae (1787) in Saint Petersburg placed a report of its correspondent Carosi (1789), a witness of the event in Kraków. Although the earthquake was known to contemporary naturalists, it was not given special attention. The earthquake was mentioned in all XIX century papers dealing with earthquake occurrence in Central Europe, i.e., Hoff (1841), Perrey (1846), Mallet and Mallet (1858), Schmidt (1858), Jeitteles (1860), Leonhard and Voltz (1896), Laska (1902) and others.

Epicenter area of the Outer Western Carpathians earthquakes in the years 1785–1786 is presented in Fig. 1.1, on the tectonic schema map of Central Europe by Guterch et al. (2007).

Whereas the earthquakes of August 22, 1785 and February 27, 1786, were located in the area bearing the geographical names of Cieszyn Silesia—Upper Silesia and Silesian Beskids, the epicenter of the December 3, 1786 earthquake has been ambiguous. The aim of this paper is to reinterpret the December 3, 1786, earthquake using known and newly found documents.

1.2 Seismicity of Outer Western Carpathians in 1785–1786

Seismicity of the Outer Western Carpathians, northwest of the Pieniny Klippen Belt, bearing geographical name of Silesian Beskids, is poorly recognized due to very low frequency of earthquake occurrence. According to the present knowledge, the region was exceptionally seismically active in the two-year period 1785–1786. The main earthquakes of years 1785–1786 are exceptional seismic events, historically known and instrumentally recorded, as concerns their magnitudes and depths of epicenters, reaching $M_W = 5.3$ and $h = 35$ km, respectively. Even quakes of magnitude $M < 2$ have not been observed there nowadays by local seismic stations. Depths of foci in the Western Carpathians are shallower, up to 20 km, but mostly not exceeding 10 km (Procházková and Dudek 1980).

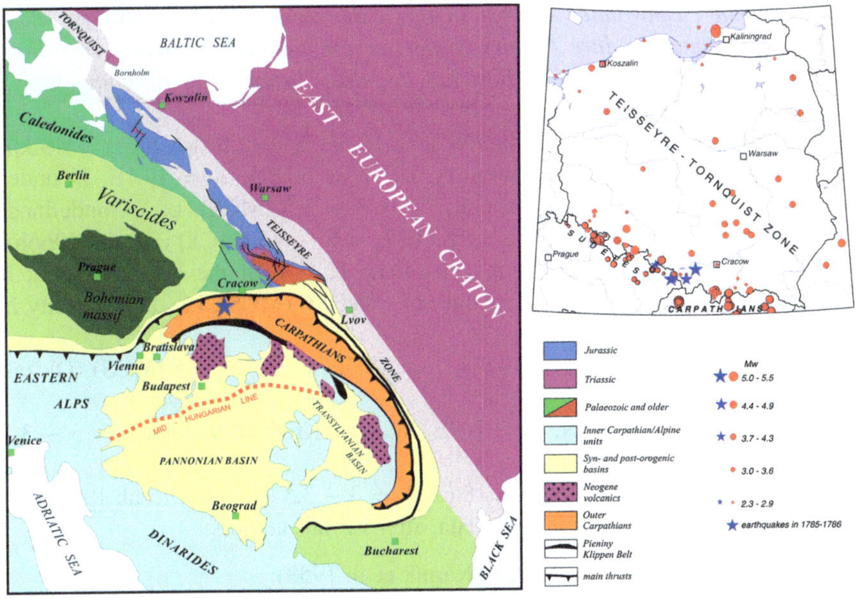

Fig. 1.1 Tectonic sketch map of Central Europe by Guterch et al. (2007) and epicenter region (marked by *asterisk*) of the Outer Western Carpathians earthquakes in the years 1785–1786. *right upper corner*—earthquake epicenters in Poland in the Mw scale since the XVI century, epicenter location after Guterch (2009)

The following earthquakes were recorded in the Outer Western Carpathians and Sudetes in the years 1785–1786:

1785 February 07 *Strong quakes in Jihlava* (49.40N, 15.59E), *Mor. Krumlov* (49.05N, 16.32E) *and in other places in Moravia and Silesia [Starke Erschütterungen in Jihlava, Mor. Krumlov und an anderen Orten Mahrens und Schlesiens]* (Kárník et al. 1958); epicenter data undefined,
1785 August 22, $\varphi = 49.7$N, $\lambda = 19.0$E, $I_0 = 6.5$ (MCS), h = 35 km (Olczak 1965); $I_0 = 6 - 7$ (EMS), h = 16 km (Guterch 2009),
1786 February 13, 24 h, Eastern Sudetes, $\varphi = 50.0$N, $\lambda = 17.5$E, $I_0 = 5.5$ (MCS) (Olczak 1966b); $\varphi = 50.15$N, $\lambda = 17.25$E, $I_0 = 5$ (EMS) epicenter estimated in this paper after data given by Remeš (1930). Remeš (1930) has stated after Brünner Zeitung (1786), no. 15, pp. 118–119: *Brno, February 21, On February 13 a torrential earthquake was felt in the area of Freudenthal (*Bruntál 49.98N, 17.45 E*), especially in localities Altstadt* (Staré Město 50.17N, 16.96E), *Schreiberseifen* (Scrbovice—a village near Bruntál) *and Dittersdorf* (Dytmarów 50.16N, 17.66E). *It happened about midnight, beds were shaking so that the sleeping people woke up, escaped their rooms and many run outdoor in panic. It was reported that the quake lasted for about* 2 min. *[Brün, den 21 Februar, Den 13. Dieses wurde in der Gegend von Freudenthal, besonders in den Ortschaften Altstadt, Schreiberseifen und Dittersdorf ein sehr heftiger Erdstoss verspürt. Es war um Mitternacht und

die schlafenden Einwohner wurden in ihren Betten dergestalt erschüttert, dass sie darüber erwachten, ihre Zimmer verliessen und viele von Schrecken aus ihren Häusern entliefen. Man setzt im Berichte, bei, dass die Erschütterung ungefähr 2 Minuten angehalten habe]
1786, February 27, 04 h, $\varphi = 49.7$N, $\lambda = 18.5$E, $I_o = 7.5$ (MCS), h = 40 km (Olczak 1966b) **Foreshocks: February 10**, $\varphi = 50.2$N, $\lambda = 19.1$E, I_o undefined; **February 26**, at night, 2 shocks, $I_o = 3 - 3.5$ (MCS), φ, λ—undefined; **February 27**, 01 h, $\varphi = 50.0$N, $\lambda = 18.0$E, $I_o = 6 - 6.5$ (MCS) (Olczak 1966b). Macroseismic data are insufficient to estimate $I_o \geq 6$ of the last foreshock; it seems that macroseismic intensities did not exceed 6 (MCS), i.e., the intensity I_o would be 5–6 (EMS).
1786, February 27, 18 h, $\varphi = 49.7$N, $\lambda = 18.5$E, $I_o = 3 - 3.5$ (MCS) (Olczak 1966b);
1786, March 04, $\varphi = 50.5$N, $\lambda = 16.7$E (Sudetes, Kłodzko Valley), $I_o = 3 - 3.5$ (MCS), (Olczak 1966b)
1786, October 03, $\varphi = 49.7$N, $\lambda = 18.6$E, $I_o = 3 - 3.5$ (MCS) (Olczak 1966b)
1786, December 03, 17 h, epicenter data, after various authors:

$\varphi = 50.1$N, $\lambda = 18.8$E, $I_o = 8$ MCS (Kárník et al. 1958)
$\varphi = 49.7$N, $\lambda = 20.0$E, $I_o = 8$? MCS (Kárník 1961)
$\varphi = 49.7$N, $\lambda = 20.0$E, h = 40, $I_o = 7.5$ MCS (Olczak 1966a)
$\varphi = 49.9$N, $\lambda = 19.3$E h = 35 $I_o = 7$ EMS (Guterch and Lewandowska-Marciniak 2002)

The earthquake of December 03, 1786, completed the set of events in the years 1785–1786 (see Fig. 1.12, Sect. 1.8).

Earthquakes recorded in the Silesian Beskids, Cieszyn Silesia and Upper Silesia since the XVIII century (except of the events of the years 1785–1786 mentioned above) are the following:

1715 or 1716, May 1, $\varphi = 49.8$, $\lambda = 18.6$, $I_o = 6$ (MSK) (Schenková et al. 1999). *A strong earthquake was felt in Schlesisch Cieszyn County, which lasted for about* 36 hours, *mentioned is Freystädtel (Schlesien)* (Fryštát 49.85N, 18.54E) *[Im schlesischen Fürstenthum Teschen wurde ein starkes Erdbeben, das '36 Stunde lang dauerte, gespürt', genannt wird Freystädtel (Schlesien)]* (Kárník et al. 1958)
1855, January 15, $\varphi = 49.8$N, $\lambda = 18.8$E, $I_o = 4$ EMS (Guterch and Lewandowska-Marciniak 2002); *Earthquake in Tesin?, Bielitz and Szliac Schlesien [Erdbeben in Tesin?, Bielitz und Szliac Schlesien]* (Kárník et al. 1958)
1872, December 26, $\varphi = 49.8$, $\lambda = 19.0$, $I_o = 4$ (Guterch and Lewandowska-Marciniak 2002); *Earthquake in Bilsko (Schlesien) [Erderschütterung in Bilsko (Schlesien)]* (Kárník et al. 1958)
1876, July 12, $\varphi = 49.8$, $\lambda = 18.6$, $I_o = 4$ EMS (Guterch and Lewandowska-Marciniak 2002) estimated after data reported by Pohludka (1897)
1928, May 15, 02 h 44 m, $\varphi = 49.90$, $\lambda = 18.36$, $I_o = 3$ MSK (Schenková et al. 1999); *Bohumín 3° (Mähr. Schlesien)* (Kárník et al. 1958)
1928, May 18, 01 h 20 m, $\varphi = 49.90$, $\lambda = 18.36$, $I_o = 4$ (Schenková et al. 1999); "*Bohumín 4° (Mähr. Schlesien*" (Kárník et al. 1958)

1994, **May 12**, 23 h 23 m 25 ± 2.5 s, $\varphi = 49.59 \pm 0.230$, $\lambda = 18.54 \pm 0.190$, $M_W = 2.4$ (Draber et al. 2000)
1994, **November 09**, 20 h 10 m 25 ± 1.6 s, $\varphi = 49.74 \pm 0.260$, $\lambda = 18.62 \pm 0.304$, $M_W = 2.6$ (Draber et al. 2000)

There are scanty macroseismic data concerning the historical events; none event has a macroseismic map. Thus, epicenter parameters have been assessed with small accuracy.

Prior to the XVIII century, no earthquakes have been recognized in the Outer Western Carpathians, but it should be noted that "The unified catalogue of earthquakes in central, northern and northwestern Europe" by Grünthal et al. (2009) has given for an event of January 31, 1259, the epicenter data: $\varphi = 49.9$N, $\lambda = 19.25$E, $I_0 = 7$, $M_W = 5.0$, with reference to Pagaczewski (1972). This epicenter location would point to the Outer Western Carpathians/Silesian Beskids, where the earthquakes in the years 1785–1786 occurred. These epicenter data seem to be unreliable, as Pagaczewski's (1972) note '*Destructive earthquake in Poland, particularly in Kraków. Felt in Silesia and Moravia (Brno)*', compiled after Laska (1902), Kárník et al. (1958), and Jeitteles (1860) is not sufficient to locate the epicenter. In XV century sources, there are mentions about earthquakes in Poland in the years 1257–1259, i.e., in chronicles by Długosz (1415–1480) and Miechowita (Maciej z Miechowa, ca 1457–1523). In both chronicles there are only a few references to earthquakes, too enigmatic to determine epicentral region. The short Długosz's entry (quoted from the edition of 1873, as Długosz 1873, **XI** p. 365) for the year 1257: *Pridie Calendas Februarii Cracoviae et in universis Polonorum urbibus et regionibus grandis terrae motus, apud Polonos rei huiuscemodi insuetos, pro monstro et religione habitus, editus est Tertiarum hora [On January 31, at* hora Tertiarum, *in Kraków and all towns and regions of Poland, a great earthquake occurred, to evoke an astonishment for the inhabitants quite unaccustomed to such events]* let us assume that the epicenter might have been in Poland. The note concerning the same event by Miechowita (quoted from fascimile as Miechowita 1986, p. 157) is even shorter and repeats Długosz's information: *Anno domini 1257 pridie kalendas Februarii hora terciarum, gradis terae motus in regno Poloniae [In the year of Lord 1254, on the eve of the* Kalendas Februarii (i.e., *January 31*), *at* hora terciarum, *a great earthquake occurred in the area of the Kingdom of Poland].*

Older middle age sources known to Długosz have given for the year 1257 the following information: *Anno domini 1257 terre motus magnus Polonie, Bohemie, Ungarie, Russie, Lancicie etc.* [*In the year of Lord 1257, a great earthquake occurred in Poland, Bohemia, Hungary, Russia, Łęczyca, etc.**[1]] (Rocznik Krasińskich 1878, p. 132). *January 31: Terre motus est una hora, et civitas Cracoviensis theutunico iure locatur* [*The earth was trembling for one hour, and the town of Kraków is founded under German law**] (Rocznik Małopolski 1878a,

[1] The quotations marked with an asterisk have been translated into English by Anna Dziembowska, based on the translation into Polish by Dominika Budzanowska.

kod. Kuropatnickiego (XV–XVI century), p. 168). *Terre motus factus est magnus est civitas Cracoviensis locata est teutonico iure* [*A great earthquake occurred, the town of Kraków was founded upon the German law**] (Rocznik Małopolski 1878b, kod. Szamotulski (1471), p. 169). *Eodem anno pridie Kalendas Februarii luna quarta, infra horam prime factus est terre motus* [*In the same year, on the eve of the Kalendas of February (i.e., January 31), when the Moon was in its fourth quarter, an earthquake occurred before the sunrise**] (Annales Ecclesiae Cracoviensis 1978, p. 87).

The last note let Laska (1902), professor of astronomy at the Prague University, conclude that it should be the year 1259 instead of 1257, as exactly on January 31, 1259, the moon was in *luna quarta.*

Silesian middle age sources would support this statement: *Anno Domini 1259 terre motus fuit magnus 2 Kalendas Februarii [In the year of Lord 1259, a great earthquake occurred two days before the* Kalendas Februarii (i.e., January 31)**]* (Rocznik Wrocławski Dawny 1878, p. 682) and *Anno Domini 1259 fuit terre motus magnus, et erat 2 Kalendas Februarii [In the year of Lord 1259, a great earthquake occurred, which happened 2 days before the* Kalendas Februarii (i.e., January 31)**]* (Rocznik Magistratu Wrocławskiego 1878, p. 682).

There are also other entries for the year 1258: by Długosz (quoted from the edition on 1873, as: Długosz 1873, **XI**, p. 369): *Anno eodem motus terrae per universam Poloniam exortus est ... Idem quoque motus terrae in Bohemia, Russia et Hungaria fuerat [In the same year there was an earthquake in the whole Poland.... The same earthquake has earlier occurred in Bohemia, Russia, and Hungary as well.*]* and by Miechowita (quoted from fascimile: Miechowita 1986, p. 159) *Anno quo supra, In regnis Poloniae, Hungariae, Bohemiae, & Russiae, motus terrae uniuersaliter apparuit & fuit [In the same year as before, in the whole Kingdoms of Poland, Hungary, Bohemia and Russia, the earthquake occurred all over*]* Later on, Maciej Stryjkowski (ca. 1547–ca. 1582; quoted from fascimile: Stryjkowski 1980), wrote: *Roku 1258, a Miecho kładzie 7, miesiąca lutego godziny trzeciej na dzień, straszne i okrutne i niesłychane ani zwykłe trzęsienie ziemi w Polszcze było, z wielkim strachem wszystkich ludzi...[In the year 1258, Miecho (i.e., Miechowita) wrote: on 7 February, at hour three, a horrible, dreadful, outrageous and unprecedented earthquake occurred in Poland, making people very fearful*[2]*]*. This specified a more detailed date for the same year 1258, i.e., February 7, hour 03 (canonic hour 03 means 09–12 a.m.), with reference to Miechowita, while Miechowita (cited above) did not mention the details of the date. It is the same hour of a day as that on January 31, 1257, *Tertiarum hora*, given by Długosz and for an event of February 07, 1258, given by Stryjkowski.

Thus, an earthquake of 1258 seems to be questionable. All the older, middle age sources cited above have mentioned only one earthquake, either in 1257 or in 1259, all of them on the same day, January 31. It could be referred to the same event of January 31, 1259. This might have been one of the strongest earthquakes

[2] Translated by Anna Dziembowska.

historically recorded in Poland, as macroseismic area comprised Ukraine, Bohemia, Hungary and Poland. Rocznik Krasińskich (1878) has mentioned *Lancicie*, i.e., Łęczyca (52.03N, 19.12E), administrative and ecclesial center of central Poland in those times. Presumably, the epicenter of the January 31, 1259 earthquake was located in the Western Carpathians but any precise locating would be only speculation.

1.3 The Earthquake of December 3, 1786—Review of Previous Data

Sieberg (1940) and later Kárník et al. (1958) located the epicenter of the December 3, 1786, earthquake "near Kraków" but the epicenter coordinates Kárník specified (50.1N, 18.8E) would point to the Upper Silesian region. Réthly (1952) supposed that the epicenter was farther away to the south, in Slovakia. Réthly based his assumption on the information about some unspecified damage to buildings in two localities, Spišská Nová Vés and Spišské Podhradie in Slovakia. Kárník supposed that there could be even another shock in Slovakia.

The analysis of historical data of the earthquake, with the compilation of its macroseismic map, was given by Olczak (1966a). Olczak based his map on contemporary information mainly in Gazeta Warszawska (1786), Schlesische Privilegirte Zeitung (1786), Schlesische Provinzialblätter (1786), Carosi (1789), Pamiętnik Historyczno Polityczny (1787) and later published notes, for example, Wojciechowski (1895), Grabowski (1900), Krzyżanowski (1900) and Koláček (1916). Olczak (1966a) has presented his critical opinion about the former estimation of the epicentral area of the earthquake, given in the catalogues of Sieberg (1940), Kárník et al. (1958) and Réthly (1952).

Olczak (1965, 1966b) compiled also macroseismic maps of the August 22, 1785 and February 27, 1786 earthquakes. Olczak's maps for the earthquakes in the years 1785–1786 were later revised and presented in the MSK-64 scale in Atlas of Isoseismal Maps, Central and Eastern Europe (Procházková and Kárník 1978).

1.4 A New Macroseismic Evaluation of the December 3, 1786, Earthquake

It seemed possible that in Poland, for the event so rare as the December 3, 1786, earthquake, some written notes should have survived in municipal and church archives, at least in the regions where building damage was mentioned. Thus, special searches were carried out in church, as well as in civil, archives, to find every possible piece of information about the earthquake. Specific attention was focused on parish archives as particularly reliable sources of data. The searches in archives have been brought 26 unknown documents. These manuscripts, written mostly in

Polish and a few in Latin and German, are presented in original languages in a separate chapter of this book (Guterch et al. 2015, this issue), without any corrections of orthography and grammar. A few church documents (Kraków AKMK-1, manuscript, 1786, and Kraków AKMK-2, manuscript, 1786) have been of poor quality for macroseimic studies, but because they have reflected general mood of people anxiety, caused by unknown phenomena, they have been presented too.

Assessment of macroseismic intensities of historical earthquakes based on documents of various origins and seismological importance needs an especially critical approach to the data (Gutdeutsch et al. 1987; Eisinger et al. 1992; and Grünthal 1992). Every information should be evaluated according to its historical value and macroseismic significance.

Information has been classified into 3 categories (**A**, **B**, and **C**) according to source significance Q and into 4 categories (**a**, **b**, **c**, and **d**) according to macroseismic importance q, after Grünthal (1992) with a slight modification of Grünthal's differentiation.

Source significance:

A contemporary official municipal and church documents

- reports of eyewitnesses
- reports in contemporary journals published immediately after the earthquake, relied on eyewitness reports

B short messages in contemporary journals and documents

- contemporary summaries of the earthquakes based on other people's descriptions

C papers published since the XIX century, without verifiable reference

- seismic catalogues

Quality of macroseismic information

a detailed description of local macroseismic effects (good for an intensity estimation)
b sufficient description to allow satisfactory intensity evaluation
c scanty description to allow an unsure intensity evaluation
d only information that the earthquake was felt stronger or weaker than in another locality for which some macroseismic effects were already known

- very scanty information such as "strongly felt", "weakly felt"

It was often necessary to verify the reliability of the data for different localities within the same information source. For example, Carosi (1789), "Capitaine au service du Roi et de la république de Pologne", an eyewitness of the events in Kraków, gave in his report to the Academiae Scientiarum Imperialis Petropolitanae in Petersburg, dated on January 17, 1787, detailed and reliable description of how the earthquake was felt by people in various circumstances. He also mentioned a few localities where the earthquake was felt without giving any details. Therefore, Carosi's data have been placed in category **A** for Kraków, but in category **B** for Kielce, Końskie, Pińczów, Małogoszcz, Myślenice, and Wieliczka. Similarly, for journals, a careful approach in choosing the right category was needed. This applied mainly to two journals, Gazeta Warszawska edited in Warsaw, and Schlesische Priviligirte Zeitung edited in Wrocław.

Thus, for a given locality, every source and every piece of information in this source, were separately verified. The information of the highest quality values in both categories was chosen for intensity estimation. References of the earthquake in manuscripts and contemporary journals are included, in the original languages, in a separate chapter of this book (Guterch et al. 2015, this issue). Catalogue of all localities where the earthquake was felt, including quotations translated into English, has also been presented there.

The list of all localities with their coordinates, intensities in the EMS-98 and MSK64 scale, and quality values in both categories, are presented in Table 1.1. The list includes also names of localities where the earthquake was felt of unknown intensities. About half of the localities, mainly small villages, have only a single source of information. Distribution of the number of sources for all localities where the December 3, 1786 was recorded is presented in Table 1.2.

Table 1.1 Intensity data points (IDP's) and quality data of the December 3, 1786 earthquake

No.	Locality	Coordinates		Intensity		Quality data*		
		φ	λ	MSK	EMS	Q	q	IQF
1	Barczków	50° 08′	20° 33′	5.5	5–6	A	c	2
2	Balice	50° 05′	19° 47′	6	5–7	A	d	3
3	Bielsko–Biała	49° 49	19° 03′	5	5	A	c	2
4	Bieńkowice	50° 06′	20° 29′	5.5	5–6	A	c	2
5	Bierutów	51° 07′	17° 33′	4.5	4–5	B	d	3
6	Bochnia	49° 58′	20° 26′	6	5–6	A	c	2
7	Bogucice	50° 03′	20° 29′	5.5	5–6	A	c	2
8	Bratucice	50° 04′	20° 32′	5.5	5–6	A	c	2
9	Brudzów	50° 43′	20° 40′	Felt		B		
Aftershock on Dec.09, 9 a.m.				4.5	4	B	b	1
10	Brzeg	50° 52′	17° 28′	5	5	B	d	3
11	Busko	50° 28′	20° 43′	6	6	B	c	2
12	Byczyna	51° 07′	18° 13′	4	4	B	d	3
13	Bytom	50° 21′	18° 55′	7.5	7	B	d	3
14	Cerekiew	50° 05′	20° 30′	5.5	5–6	A	c	2
15	Chęciny	50° 48′	20° 28′	Felt		B		
16	Cieszyn	49° 45′	18° 38′	6	5–6	A	c	2
17	Dobroszyce	51° 16′	17° 20′	4	4	B	c	2
18	Drogomyśl	49° 52′	18° 46′	6	5–6	B	c	2
19	Drzewica	51° 27′	20° 28′	Not felt		B		
20	Garnek	50° 53′	19° 27′	6	6	A	b	1

(continued)

Table 1.1 (continued)

No.	Locality	Coordinates		Intensity		Quality data*		
		φ	λ	MSK	EMS	Q	q	IQF
21	Głębowice	49° 57′	19° 20′	Felt		B		
22	Głogów	51° 40′	16° 05′	3.5	3–4	B	c	2
23	Głogówek	50° 21′	17° 52′	5	5	B	d	3
24	Głubczyce	50° 12′	17° 50′	5	5	B	d	3
25	Gnojno	50° 36′	20° 51′	5.5	5–6	A	c	2
26	Gorzów Śl.	51° 05′	18° 26′	4	4	B	d	3
27	Gręboszów	50° 15′	20° 47′	5	4–5	B	c	2
28	Iwanowice	51° 39′	18° 19′	5	4–5	B	c	2
29	Jaksice	50° 09′	20° 30′	Felt		A		
30	Janów	50° 43′	19° 26′	5	4–5	A	c	2
31	Kalisz	51° 46′	18° 05′	4	4	A	c	0
32	Kalwaria	49° 52′	19° 41′	6	5–6	B	c	2
33	Karwina	49° 51′	18° 33′	6	6	C	d	3
34	Kęty	49° 53′	19° 13′	6	6	A	a	0
35	Kielce	50° 52′	20° 38′	Felt		B		
36	Kije	50° 37′	20° 34′	Felt		B		
37	Kluczbork	50° 58′	18° 13′	4	4	B	d	3
38	Końskie	51° 11′	20° 24′	Not felt		B		
39	Koprzywnica	50° 36′	21° 35′	4.5	4	A	c	2
40	Korytnica	50° 40′	20° 31′	4.5	4–5	A	c	2
41	Košice	48° 43′	21° 15′	Felt		B		
42	Kowala	50° 11′	20° 20′	5.5	5–6	B	b	1
43	Koźle	50° 20′	18° 08′	5	5	B	d	3
44	Kraków	50° 04′	19° 56′	6	5–6	A	a	0
45	Krapkowice	50° 29′	17° 58′	5	5	B	d	3
46	Kromołów	50° 30′	19° 30′	6	5–6	A	a	0
47	Krzeszowice	50° 08′	19° 38′	5.5	5–6	B	c	2
48	Kurozwęki	50° 35′	21° 06′	5	5	A	c	2
49	Legnica	51° 12′	16° 10′	3.5	3–4	B	c	2
50	Lewin Brzeski	50° 45′	17° 37′	5	5	B	d	3
51	Lipowiec	50° 05′	19° 27′	Felt		B		
52	Lisów	50° 42′	20° 40′	3.5	3–4	A	b	1
53	Litovel	49° 42′	17° 05′	6	5–6	B	c	2
54	Lubsza	50° 37′	19° 00′	Felt		B		
55	Lwów	49° 50′	24° 02′	3	3	B	a	1
56	Małogoszcz	50° 49′	20° 16′	Felt		B		
57	Milicz	51° 32′	17° 16′	4	4	B	c	2
58	Mogiła	50° 04′	20° 04′	5.5	5	A	b	1
59	Mrzygłód	50° 32′	19° 22′	6	6	A	a	0
60	Myślenice	49° 50′	19° 56′	5.5	5–6	A	c	2

(continued)

Table 1.1 (continued)

No.	Locality	Coordinates		Intensity		Quality data*		
		φ	λ	MSK	EMS	Q	q	IQF
61	Nadzów	50° 11′	20° 19′	Felt		B		
62	Namysłów	51° 05′	17° 43′	6	5–6	A	a	0
63	Nowa Góra	50° 11′	19° 35′	Felt		B		
64	Nowy Sącz	49° 37′	20° 43′	4.5	4–5	A	c	2
65	Nowy Targ	49° 29′	20° 02′	7.5	7	A	a	0
66	Nysa	50° 29′	17° 20′	5.5	5	B	d	3
67	Oleśno	50° 53′	18° 26′	4	4	B	d	3
68	Oleśnica	51° 13′	17° 23′	4	4	B	d	3
69	Olkusz	50° 17′	19° 34′	5.5	5	A	c	2
70	Olomouc	49° 36′	17° 15′	5	5	B	c	2
71	Opatów	50° 48′	21° 25′	Felt		B		
72	Opava	49° 56′	17° 54′	Felt		C		
73	Opole	50° 40′	17° 55′	5	5	B	d	3
74	Oświęcim	50° 02′	19° 12′	Felt		B		
75	Pinczów	50° 32′	20° 31′	5.5	5	A	c	2
76	Piotrków	51° 25’	19° 42′	5	5	A	a	0
77	Pokój	50° 55′	17° 51′	6	5	B	c	2
78	Polanka	49° 52′	19° 56′	5	5	A	a	0
79	Popędzyna	50° 08′	20° 32′	5.5	5–6	A	c	2
80	Potok Złoty	50° 42′	19° 26′	Felt				
81	Prudnik	50° 19′	17° 35′	5	4–5	B	d	3
82	Pszczyna	49° 59′	18° 57′	6	6	A	a	0
83	Puławy	51° 25′	21° 58′	Felt		B		
84	Racibórz	50° 05′	18° 13′	5.5	5	A	b	1
85	Racławice	50° 19′	20° 14′	6	6	B	c	2
86	Radoszyce	51° 05′	20° 16′	Not felt		B		
87	Rawicz	51° 37′	16° 52′	4	4	B	c	2
88	Rybna	50° 28′	18° 48′	7.5	7	B	d	3
89	Rybnik	50° 06′	18° 33′	5.5	5	B	c	2
90	Sancygniów	50° 25′	20° 18′	5.5	5	B	c	2
91	Sandomierz	50° 41′	21° 45′	5.5	5	A	b	2
92	Spišska Nová Vés	48° 57′	20° 34′	6	5	B	c	1
93	Spišske Podhradie	49° 00′	20° 45′	Felt		B		
94	Stolec	50° 35′	16° 53′	5	5	B	d	3
95	Sułów	51° 30′	17° 10′	4	4	B	c	2
96	Świerklaniec	50° 26′	18° 56′	7.5	7	B	d	3
97	Tarnowskie Góry	50° 27′	18° 51′	7	7	A	a	0
98	Tarnów	50° 01′	20° 59′	6	6	A	a	0
99	Trzebinia	50° 09′	19° 28′	Felt		B		
100	Trzebnica	51° 18′	17° 03′	4	4	B	c	2

(continued)

Table 1.1 (continued)

No.	Locality	Coordinates		Intensity		Quality data*		
		φ	λ	MSK	EMS	Q	q	IQF
101	Twardogóra	51° 22′	17° 28′	4	4	B	c	2
102	Tyniec	50° 01′	19° 48′	6	5–6	B	c	2
103	Ujście Solne	50° 07′	20° 31′	5.5	5–6	A	b	1
104	Wawrzeńczyce	50° 07′	20° 19′	Felt		B		
105	Widuchowa	50° 29′	20° 48′	5.5	5–6	A	c	2
106	Wieliczka	49° 59′	20° 04′	5	5	B	b	1
107	Wielkanoc	50° 20′	19° 55′	5.5	5	A	c	2
108	Wiślica	50° 22′	20° 41′	Felt		B		
109	Włoszczowa	50° 51′	19° 58′	6	6	A	b	1
110	Wolbrom	50° 24′	19° 45′	Felt		B		
111	Wołczyn	51° 01′	18° 03′	4	4	B	d	3
112	Wrocimowice	50° 18′	20° 15′	6.5	6	B	c	2
113	Wrocław	51° 07′	17° 02′	4	4	A	b	1
114	Wrzępia	50° 05′	20° 32′	5.5	5–6	A	c	2
115	Zator	50° 00′	19° 26′	Felt		B		
Aftershock on Dec.04, 3 a.m.				3.5	3	A	b	1
116	Zduny	51° 39′	17° 23′	4	4	B	c	2
117	Ziębice	50° 36′	17° 02′	5	5	B	d	3
118	Zębocin	50° 10′	20° 19′	6	6	B	c	2
119	Žilina	49° 13.4′	18° 44.4′	Felt		D		

*Q is the quality of source significance, q the quality of macroseismic significance, and IQF the intensity quality factor

Table 1.2 Frequency of sources for localities where the December 3, 1786 earthquake was recorded

Number of localities	56	25	11	6	7	4	1	1	1
Number of sources	1	2	3	4	5	6	7	10	26

1.4.1 Macroseismic Intensity Assessment for Kraków

The greatest number of written sources about the earthquake was available for Kraków, the capital of the province in southern Poland. Data set comprises 26 references for Kraków including 10 of quality Q = A and q = a. This relatively large amount of data for one locality tends to exaggerate the local intensity.

In the Kraków archives of the Augustian Friars (Kraków AP-1, manuscript, 1786) and Acta Actorum Capituli (Kraków AKK, manuscript, 1786-1787) there is confirmation of some damage to the vaulting of St. Catherine's Church of the

Augustian Friars; the same information is reported by Gazeta Warszawska (1786, Dec. 16). The relevant excerpts are as follows:

> *…Because of the earthquake the vaulting of our church cracked in six spots over chancel, disintegrating with great danger, which was quickly repaired…* (Kraków AP-1, manuscript, 1786),
>
> *…A greater shake could be felt in St. Catherine's Church of the Augustian Friars, where the vaulting broke, opening wide and threatening to callapse…* (Kraków AKK, manuscript, 1786–1787)
>
> *…The huge and high church of the Augustian Friars got so cracked in the roof of its chancel that it had to be closed for safety reasons…*(Gazeta Warszawska 1786, Dec. 16).

The presbytery of this great gothic church of St. Catherine's built in the year 1378 is exceptional; it is 26 m high, while only 10.5 m wide, and this was the second time that the vaulting was damaged by the earthquake. The vaulting of the same church collapsed during the earthquake of 5 June 1443, with its epicenter in Central Slovakia. Długosz (quoted as Długosz 1877, XIII, p. 691), the witness of the event, has written: *Quinta mensis Iunii generalis terrae motus praesertim in Poloniae, Hungariae et Bohemiae Regnis et partibus vicinis, adeo validus exortus est, ut turres, aedificiaque murorum corruerent, et singulae domus quantumcunque robustae aut firmae, motu notabili volverentur, fluviorum alvei, aquis In partes utrasque diffugientibus, vacui cernerentur, liquida quaequae salirent, hominess pavore subito consternati a* sensu *et ratione alienarentur. Testudo monasterii Sanctae Catharinae fratrum Beati Augustini in Casimiria, motu illo in terram nocte decidit, et plura alia loca motu terrae ruinata sunt. Intensior tamen motus ipse in Regno Hungariae fuit, ubi et castra quardam eversa sunt' [On the day 5 of month June, there was an earthquake in Poland, Hungary, Bohemia and adjacent countries, so violent that the turrets and brick buildings were destroyed, and the most robust structures were falling down; rivers went out of their beds and poured on both sides, leaving the bottoms dry, while waters silted up everything; people, under a sudden panic were getting out of their wits. The vault of the St. Catherine's church at the Augustinian Friars monastery in Kazimierz collapsed, as well as many other buildings. Still stronger earthquake effects were in Hungary, where even some castles collapsed.*[3]*].*

During about 1,000 years of Kraków's history, the town and its churches, the oldest witnesses of events, have never suffered serious earthquake damage except of the case of St. Catherine's church. It seems that the structure of the church vaulting was weakened and easy to be damaged. Outside views of the church, i.e., from the presbytery and nave side are presented, respectively, in Figs. 1.2 and 1.3. Inside views of the church from the nave side towards the presbytery are presented in Figs. 1.4a, b and 1.5. A recent crack of the vaulting nave above a window, and a steel support of the vaulting nave is seen in Fig. 1.5.

[3] Translated by Anna Dziembowska, based on translation into Polish by Zofia Kurzelanka.

Fig. 1.2 St. Catherine's church in Kraków, the view of the presbytery

Only few reports survived up to now about some damage to common, ordinary people dwellings in Kraków, albeit without special emphasis:

> *...different signs of destruction remained in many houses* (Kraków AKAP-2, manuscript, 1786),
>
> *...nothing collapsed anywhere, but in many places walls and ceilings cracked* (Krzyżanowski 1900),
>
> *...a few pieces of a cracked cornice fell to the ground...the earthquake caused no noticeable damage* (Carosi 1789),
>
> *...without any greater harm, apart from some small cracks* (Gazeta Warszawska 1786, Dec. 09)
>
> *...in many places walls cracked, in particular at the Monastery of the Carmelites at Piasek. In this monastery one of the friars felt very weak in the legs; but when the earth shook and the wall in his cell cracked over the door... he escaped from the cell by himself...* (Gazeta Warszawska 1786, Dec. 16).

The damage to the walls of the Carmelites Monastery seems not serious, as it has not been confirmed by church archives.

There is also evidence that the earthquake was felt differently in various parts of the town. It was mostly felt near the Vistula river at Kazimierz, a suburb of Kraków in those days (Kraków AKK manuscript, 1786-1787; Gazeta Warszawska 1786, Dec. 13; Carosi 1789). According to Kraków AKK manuscript (1786): *...statues in different buildings came down as, e.g., at Kazimierz, in the Lateran Canons Corpus Christi Church*. There, at Kazimierz, only about 300 m from the Vistula, is also located the above-mentioned St. Catherine's church. Supposedly,

Fig. 1.3 St. Catherine's church in Kraków, the view of the nave

the river deposits that were not well consolidated could have locally increase the macroseismic effects.

From recorded damage to the buildings, the intensity 6 EMS in Kraków cannot be assessed, as it would mean that 'many', i.e., 20–60 %, of houses suffered slight non-structural damage. Such a conclusion would be much exaggerated taking into account that reports might tend to underline extreme cases for such unusual event in Poland as an earthquake. This might concern mainly descriptions published in papers, but the daily journal Gazeta Warszawska (1786) seems to be most reliable and objective. We can consider that there were less than 20 % houses affected by the earthquake. Nothing is said about vulnerability class of the damaged houses. There were mainly tenements, i.e., brick houses up to 4 stores, in the center of Kraków, with timber ceilings and ruffs covered by slates, of vulnerability class A and B in the EMS scale. According to Encyklopedia Krakowa (2000), in the year 1787, there were 539 tenements. Wooden one-storied houses, of higher vulnerability class, with roofs covered by wooden slates, were mainly in the suburbs. Thus, damage to dwellings suggests the intensity 5 EMS in spite of the case of St. Catherine's church.

It is not known whether people left houses and run outside. Population of Kraków at the end of XVIII century was about 24,000, and such information should be reported by daily journals. (According to Encyklopedia Krakowa (2000), population of Kraków together with suburbs, in the year 1787, was about 21,500–23,100

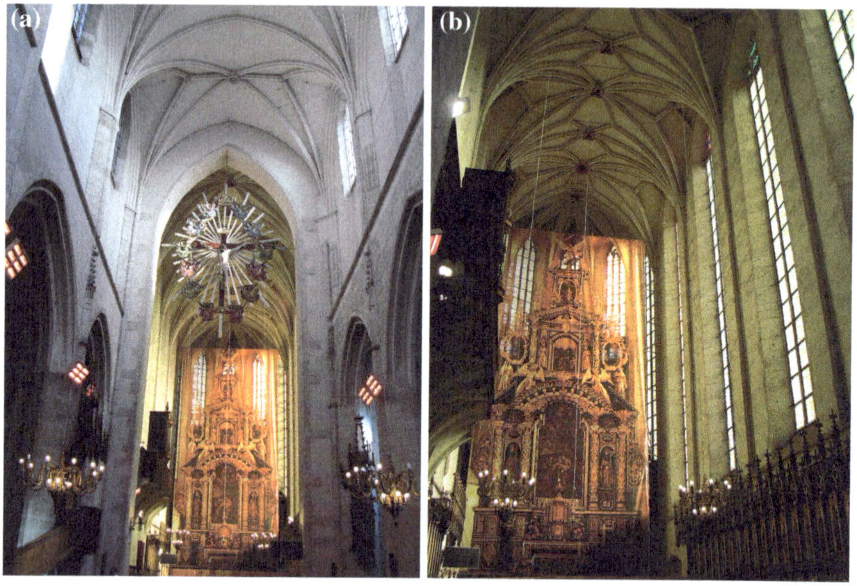

Fig. 1.4 **a**, **b** St. Catherine's church in Kraków, the inside views of the presbytery from the nave side

Fig. 1.5 St. Catherine's church in Kraków, the vaulting of the nave. A recent crack marked by an *arrow*

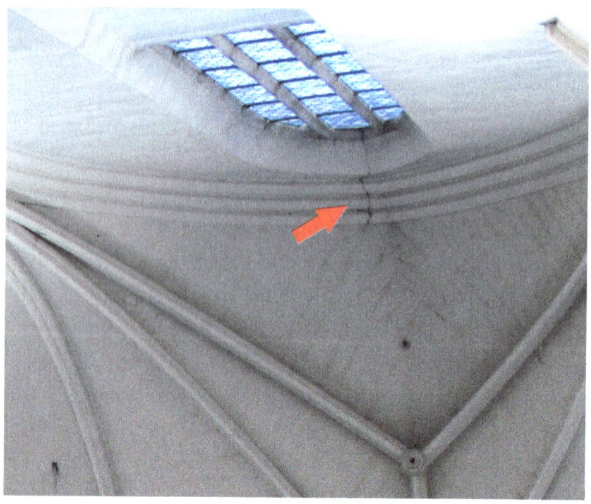

inhabitants.) There is also a lack of data whether the earthquake was felt outdoors, apart from people at the bridge. It is only Carosi (1789) who has written that standing persons found it difficult to keep balance and there were even those who fell over, but he has not specified if it happened indoors or outdoors. There is no data concerning the effects of the earthquake felt by people to confirm the intensity 6. Summarizing all the data, it can be supposed that the intensity in Kraków could reach 6 in the EMS scale in areas near Vistula, but for the whole town it is assessed at 5–6.

1.4.2 Macroseismic Intensity Assessment for Tyniec

Macroseismic intensity at Tyniec, located at a high bank of the Wisła river, about 12 km south from Kraków, was evaluated at 5–6 in the EMS scale, on the basis of contemporary information. A view of Benedictine Monastery at Tyniec from the Wisła river is presented in Fig. 1.6. There was some unspecified damage to the Benedictine Monastery walls, but the data are insufficient to assess intensity 6. Słownik Geograficzny (1892), i.e., edited 100 years after the earthquake, gave information that *the earthquake in 1786 caused some damage to walls* (Monastery walls) *and torn a stone block, that fell to the Wisła river.* There are no data in contemporary sources about this falling down of stone block, which would be too conspicuous to be unnoticed. It should even be felt nearby, while a stone block of 1,000–1,500 tons was falling 15–20 m down (Alexandrowicz 1956). According to that author, this must have happened in postglacial period, but more precise date cannot be assessed from geological data. Traces of the stone block tearing are still visible in Jurassic rock at Tyniec, right down of the Benedictine Monastery, with part of the block in the river (see Fig. 1.7). Limestone rocks are easily tractable by erosion and it seems possible that the earthquake could have triggered the process of division along one of many crevices.

Fig. 1.6 Benedictine Monastery at Tyniec

Fig. 1.7 The limestone block in Wisła river and traces of its tearing in the rock, right down of the Benedictine Monastery at Tyniec—traces of the stone tearing in the rock marked by an *arrow*

1.5 Foreshocks/Aftershocks

No foreshocks have been known over a period of one month before the main earthquake. Only two aftershocks have been recorded:

(1) **December 4**, at 03 a.m. at Zator (50.00N, 19.44E) with felt intensity of 3 EMS. At Zator, where the main earthquake intensity could be hardly estimated, there is reliable evidence of an aftershock: *...the local doctor, J.P. Gherri, an Italian, born in Bologna, knowing from the experience in his native country that such events are often not singular but may be repeated, did not sleep the whole night and testified that there was another earthquake at 3 a.m., but one which those asleep did not notice…* (Gazeta Warszawska 1786 Dec.16)

(2) **December 9**, at 08 a.m. at Brudzów (50.72N, 20.67E) with felt intensity of 4 EMS. At Brudzów, there are no data for evaluation of main earthquake, but there is confirmation of an aftershock: ... *at the Brudzów village the second time the earthquake was felt on December 9 at 8, at sunrise... hooks at doors moved, locks opened, hanging bars knocked window-panes rattled…* (Kraków AKMK-6, manuscript, 1786)

1.6 Macroseismic Maps

1.6.1 Intensity Data Points

Intensity data points (IDP's) in the EMS scale were grouped in 4 categories of different intensity quality factors (IQF) according values of source significance Q (**A**, **B**, and **C**) and macroseismic importance q (**a**, **b**, **c**, and **d**):

IQF0—IDP's of macroseismic importance **a** evaluated from **A** sources
IQF1—IDP's of macroseismic importance **a** evaluated from **B** sources and of macroseismic importance **b** evaluated from **A** and **B** sources.
IQF2—IDP's of macroseismic importance **c**.
IQF3—IDP's of macroseismic importance **d**.

IDP's of IQF0 and IQF1, i.e., of the highest quality, are presented in Fig. 1.9. All IDP's, i.e., of IQF0, IQF1, IQF2, and IQF3 are presented in Fig. 1.10. In spite of the fact that IDP's differ much in quality of data, general patterns of their distribution for the 4 quality categories are similar. Great deal of information about the earthquake follows the two main rivers, Wisła and Odra, where towns have been located. Considerable data have been available from Lower Silesia, to the NE of Wrocław, and from the surroundings of Kraków. These were the areas of highest population in those days.

The information from the Slovakia-Poland borders in the Carpathians is scanty, except of the administrative and commercial centers of the region as Nowy Targ and Nowy Sącz in Poland. Intensity 7 was evaluated from a source of highest quality for Nowy Targ, located 20 km north of the Tatra Mts. It seems that the meager populations and the lack of communication between small villages in this upland area caused that little reports about the earthquake have been written. Further to the south, in Slovakia, there are reports that the earthquake was felt in Košice, Spišská Nová Vés and Spišské Podhradie, but the data are of poor quality and only for Spišská Nová Vés it was possible to determine intensity 5 of category IQF3. There are no other data from the Inner Carpathians. There is also scanty information from the Bohemian Massif. The only reports are those from Litovel (I = 5–6) and Olomouc (I = 5), but both of IQF3.

There is a lack of reports testifying that the earthquake was felt in the Holy Cross Mts region. It was always a well populated area. In general, the earthquake was not felt to the east of the Teisseyre-Tornquist zone.

1.6.2 Isolines

A pattern of IDP's distribution of category IQF0 and IQF1 allowed to determine isoline of intensity 6 and probable course of the northern part of isoline 5 (see Fig. 1.8). It can be noted that Nowy Targ (I = 7) would be at least 50 km outside isoline 6.

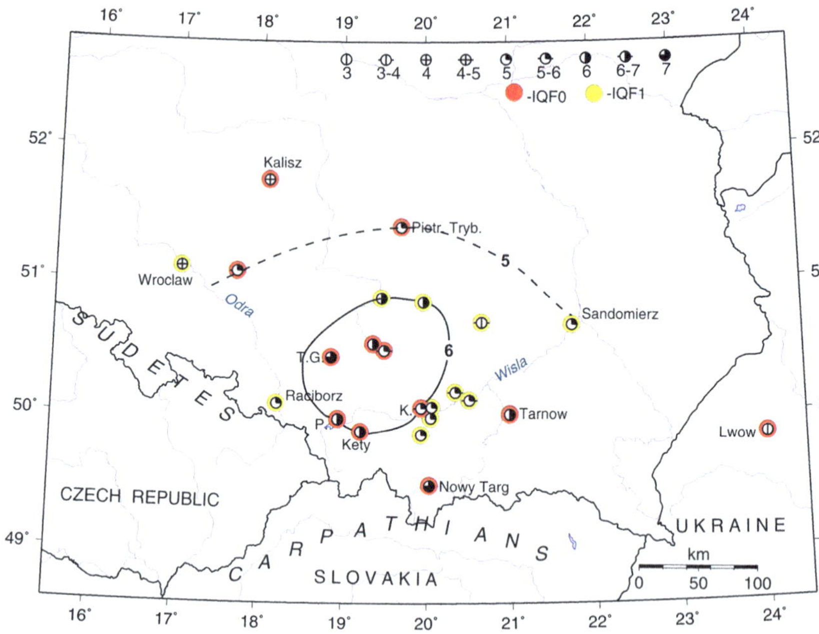

Fig. 1.8 Macroseismic map of the December 3, 1786 earthquake—most reliable intensity data points (IDP's), in the EMS scale of intensity quality factors IQF0 and IQF1, K.—Kraków, P.—Pszczyna, Piotr. Tryb.—Piotrków Trybunalski, T.G.—Tarnowskie Góry

IDP's of category IQF2 generally supplement IDP's of higher qualities from the west and south of isoline 6. This allowed to add probable course of isoline 5 and a section of isoline 4 from the northwest of the epicenter. IDP's of category IGF3 between longitudes 18 and 19E seems to be the most questionable. The quake was felt in the east up to Lwów (I = 3). Isolines of December 3, 1786, earthquake determined from all IDP's are presented in Fig. 1.9.

1.7 Parametres of the Earthquake

In spite of nearly 100 IDPs there is a problem with determining the epicenter of the earthquake of December 3, 1786. The pattern of IDP's distribution in the epicentral area is ambiguous. The highest intensities, 7, are dispersed and cannot be of any help in determining the epicenter. The epicenter should be in the area encircled by isoline of intensity 6. The size of macroseismic area points to the deep crustal epicenter.

Cecić et al. (1996) discussed various procedures used for determining epicenter coordinates from macroseismic data and suggested the barycenter as the epicenter, i.e., the point which is the center of the recorded macroseismic effects. The barycenter, determined directly from the IDP's, would be the most objective method for derivation of the earthquake coordinates.

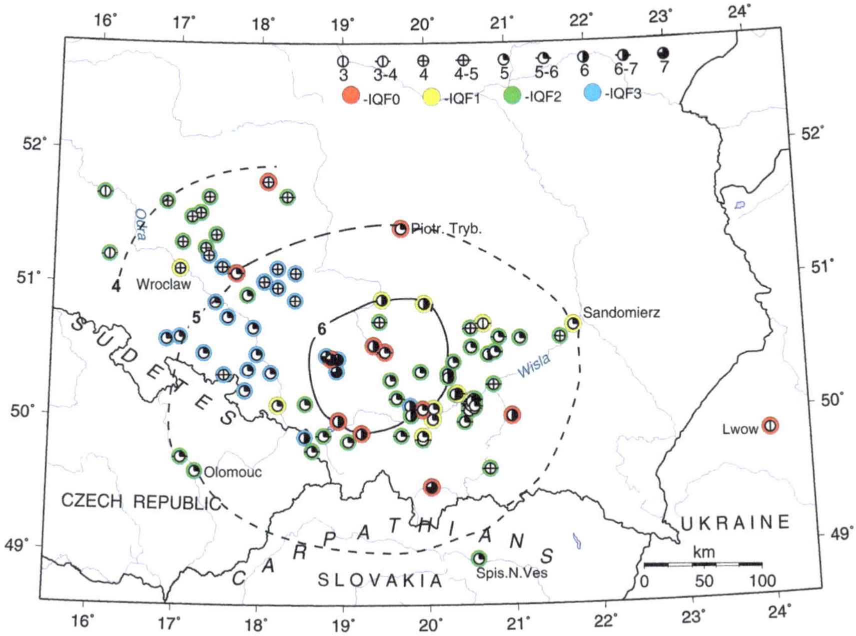

Fig. 1.9 Macroseismic map of the December 3, 1786 earthquake—all intensity data points (IDP's) in the EMS scale of intensity quality factors: IQF0, IQF1 IQF2 and IQF3. Piotr. Tryb.—Piotrków Trybunalski, Spis. N. Vés—Spišská Nová Vés

Such a method, i.e., the Macroseismic Estimation of the Earthquake Parameters (MEEP), for derivation of earthquake parameters based directly on IDP values, was presented by Musson and Jiménez (2008). The December 3, 1786, earthquake epicenter location was made with the use of software MEEP version 2.0, developed in the frame of the NA-4 module of the NERIES project by Musson (2009) in the British Geological Survey and targetted at shallow earthquakes, with $h < 20$ km. The major aim of the NA-4 module of the European Framework project NERIES was to produce a uniform catalogue of European historical earthquakes that occurred before 1900.

Of the four algorithms provided by the MEEP version-2 software, three algorithms have been used for derivation of epicenter location (φ, λ): Centroid, Pairwise (Shumila), and MEEP:

The Centroid method estimates firstly an initial centroid of the highest IDP's after Gasperini et al. (1999). Outlier data points are trimmed and the centroid of the next highest IDP's is calculated. The process continues until an adequate number of data points is available.

The Pairwise (Shumila) method is the Musson and Jiménez (2008) version of Shumila's technique, which was never published. The principle of the method is based on unquestionable fact that for any two unequal IDP's, the epicenter should be closer to the higher one. All possible pairs of unequal IDP's are examined. The more cases where the higher value is closer to the test point, the more likely the test point should be the epicenter. The Pairwise technique calculates only epicenter coordinates of the tested earthquake.

MEEP method takes the centroid epicenter as a starting point to the further steps based on a grid search routine. Epicenter location is based on the Kövesligethy relation

$$I_0 - I = k \log (r/h) + k\,\alpha \log e\,(r - h),$$

where r is the distance to hypocenter, h is the depth, k is a constant representing isoseismal spacing, usually assumed around 3, and α is an inelastic attenuation coefficient. For the given parameters k, α, and intensity I_0, the grid search routine should find the best fitting of the Kövesligethy model and the provided IDPs. The lowest misfit at the final iteration is taken to be the best epicenter (Musson and Jiménez 2008; Musson 2009). MEEP moment magnitude is based on Frankel (1994) relation.

The MEEP method was developed by Musson et al. (2008) for seismic areas where there are both, instrumentally well determined earthquakes parameters in a wide range of magnitudes, and sets of IDPs of the same events. Such data as k parameter in Kövesligethy formula, scaling factor C in Frankel's (1994) magnitude relation are needed to be found by an additional procedure, which is also used to adjust the assumed value of regional shear wave attenuation Q.

The earthquakes in the years 1785–1786 are the only seismic events historically known and instrumentally recorded in the region for which macroseismic maps are available. Even shocks of magnitude $M < 2$ have not been observed there by local seismic stations. For calibrating these parameters, the nearest seismic region would be the Orawa–Nowy Targ Basin, Western Carpathians, located at the northernmost part of the Pieniny Klippen Belt (PKB). However, tectonic patterns of these two regions are different. Depths of foci in the northern part of the PKB are shallow, up to 5 km; the absorption of seismic energy is large. Attenuation coefficient α in the Kövesligethy formula in the PKP is around 0.05 after Kárník et al. (1977); α determined for above mentioned earthquakes by Guterch et al. (2005) and Guterch (2006) is in the range $0.025 < \alpha < 0.050$.

The α value for the Western Beskids (Outer Western Carpathians) is 0.001 after Kárník et al. (1977) and 0.006 after Procházková and Dudek (1980). Procházková and Dudek (1980) got the attenuation coefficient from the Outer Western Carpathians earthquakes recorded in the years 1785–1786, based on data provided by the Atlas of Isoseismal Maps for Central and Eastern Europe (Procházková and Kárník 1978). After Kárník (1968), the α value determined for a much broader region, encompassing the Bohemian Massif, is in the range $0.005 < \alpha < 0.025$.

The α and k values are the only values besides $\Delta I_0 = I_0 - I_{max}$ (the assumed margin of the intensity increase above the observed I_{max}) that affect the location of the MEEP epicenter; these values do not affect the Centroid and Pairwise epicenter.

1.7.1 Parameters of the December 3, 1786, Earthquake

Assuming $\alpha = 0.005$, isoseismal spacing $k = 3.0$ and margin of intensity increase $\Delta I_o = 0.25$, the MEEP201 program was run for IQF thresholds 0, 1, 2, and 3. The way of ordering IQF to an individual IDP was explained above. Epicenter location obtained for thresholds 0, 1, 2, and 3 are presented in Table 1.3. Results of these 3 methods for epicenter location of December 3, 1786, earthquake are not in a satisfactory agreement taking into account uncertainties ranging from ±10 km up to ±98 km. The lowest uncertainties are for threshold 3, i.e., for all IDP's taken into calculation. All epicenters but the one got by Pairwise method for threshold 0, are located inside the isoline of intensity 6 (see Fig. 1.10).

- The Centroid epicenter location varies and depends on a level of IDP quality threshold
- The Pairwise epicenter location varies the most for thresholds 0 and 1, but it is stable for thresholds 2 and 3. This method, unreliable for small datasets, can be accurate after Musson (2009) for larger datasets. Pairwise location for threshold 0, for which there were only 11 IDP's, is the only one outside an isoline of intensity 6.
- The results of MEEP method location hardly depend on the quality threshold of IDP's. This is mainly due to the assumed margin of intensity increase $\Delta I_o = 0.25$. The pattern of macroseismic data does not allow to increase this margin. The lowest uncertainty value, ±13.2 km, has been got for threshold 3.

Taking into account that IDP's of intensity quality factors (IQF) 0 and 1 are the most reliable, the values of epicenter location are the following:

- for threshold 1 got by Centroid method: $\varphi = 50.4$N, $\lambda = 19.3$E ± 37 km
- for threshold 1 got by MEEP method: $\varphi = 50.1$N, $\lambda = 19.2$E ± 53 km
- threshold 2 got by Pairwise method: $\varphi = 49.9$N, $\lambda = 19.4$E ± 55 km

The epicenter location got by the Centroid method, i.e., $\varphi = 50.4$N, $\lambda = 19.3$E ± 37 km, was assumed to be the most objective location of December 3, 1786 earthquake.

The MEEP and Centroid methods calculate full location (φ, λ, h) that fits the best the Kövesligethy formula. As it was expected, Centroid and MEEP depth values are both 20 ± 0 km, that would mean the maximum depth calculated by the program. The depth of December 3, 1786, earthquake is evidently higher than 20 km as the size of macroseismic area points to the crustal epicenter. Thus, the depth was calculated directly from the Kövesligethy formula by Sponheuer's (1960) method.

The mean radiuses of isoseismals 6, 5 and 4 were estimated to be, respectively, 60–65, 130–140, and 190–210 km. Assuming the epicentral intensity 7.25 ($I_o = I_{max} + 0.25$) the best fit of the computed and observed data has been obtained for h = 35 km and attenuation coefficient $\alpha = 0.004$.

Table 1.3 Epicenter locations of main earthquakes in the years 1785–1786 in the Outer Western Carpathians

December 03, 1786

Method	Threshold 0: 11 IDP's			Threshold 1: 23 IDP's			Threshold 2: 69 IDP's			Threshold 3: 92 IDP's		
	φ	λ	(km)	Φ	λ	(km)	φ	λ	(km)	φ	λ	(km)
Centroid	50.146	19.662	±51.7	50.431	19.299	±37.4	50.353	19739	±39.7	50.410	18.907	±9.8
Pairwise	49.487	19.003	±78.2	50.536	19.072	±98.3	49.925	19.443	±55.3	49.925	19.438	±36.3
MEEP	50.112	19.300	±48.8	50.122	19.198	±52.6	50.184	19.138	±65.7	50.198	19.181	±13.2

August 22, 1785

Method	Threshold 0	Threshold 1	Threshold 2	Threshold 3: 18 IDP's		
Centroid				49.874	18.777	±16.5
Pairwise				49.366	19.293	±54.9
MEEP				49.734	19.084	±14.3

February 27, 1786

Method	Threshold 0	Threshold 1	Threshold 2	Threshold 3: 47 IDP's		
Centroid				49.765	18.328	±25.8
Pairwise				49.701	18.711	±23.2
MEEP				49.618	18.608	±13.8

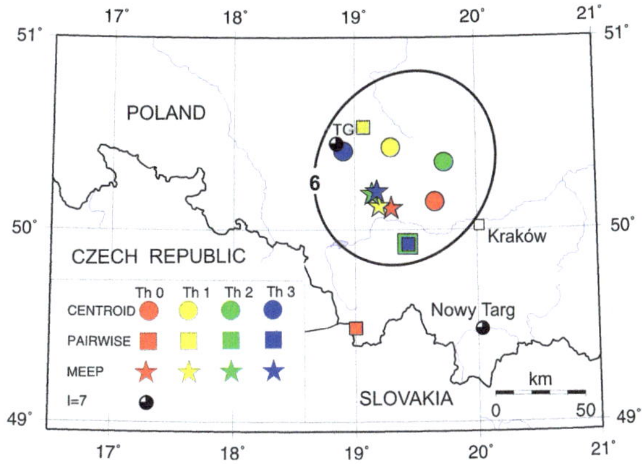

Fig. 1.10 Epicenter location of December 3, 1786 got by Centroid, Pairwise, and MEEP methods for intensity data points (IDP's) of different thresholds (Th) of intensity quality factor (IQF). Isoline of intensity 6 is presented. TG = Tarnowskie Góry

For h = 35 km and Io = 7 (EMS scale), macroseismic magnitude M_M, local magnitude M_L and moment magnitude M_w were obtained as follows:

$$M_M = 5.4, \quad M_L = 5.6, \quad \sigma = 0.39, \quad M_w = 5.3, \quad \sigma = 0.30$$

Magnitude M_M was derived after Kárník's (1968) relation for the Western Carpathians: $M_M = 0.55\ I_o + 0.93\ \log h + 0.14$. Magnitudes M_L and M_w were determined using empirical relations developed by Grünthal et al. (2009) for historical earthquakes in central, northern, and northwestern Europe:

$$M_L = 0.810\ I_o + 0.49\ \log h - 0.85, \quad M_w = 0.0376\ M_L x2 + 0.646\ M_L + 0.53.$$

1.7.2 Parameters of August 22, 1785, and February 27, 1786, Earthquakes

The same methods were used to estimate earthquake parameters of the Outer Western Carpathians earthquakes on August 22, 1785, and February 27, 1786, i.e., for epicenter location (φ, λ)—the MEEP method by Musson et al. (2008), depths were calculated directly from the Kövesligethy formula by Sponheuer (1960) method, and moment magnitudes after Grünthal et al.'s (2009) formula. Epicentral intensity $I_o = 7.25$ ($I_{max} = 7$, $\Delta I_o = 0.25$) was assumed for depth calculation of February 27, 1786 earthquake and $I_o = 6.25$ ($I_{max} = 6$, $\Delta I_o = 0.25$) for depth calculation of August 22, 1785 earthquake.

IDP's were evaluated in the EMS scale after macroseismic data presented by Olczak (1965) for the earthquake of August 22, 1785 and Olczak (1966b) for the earthquake of February 27, 1786. Macroseismic data were supplemented by information from manuscripts (Cieszyn AP, manuscript, 1767–1822; Kęty MO, manuscript, 1785-1795) and magazines (Gazeta Warszawska 1786, Dec 13, No. 99; Lwowskie Tygodniowe Wiadomości 1786, Dec 23, No. 95; Brünner Zeitung 1786, No. 101 and No. 104: Schlesische Provinzialblätter 1786, vol. 4, No. 12, Schlesische Privilegirte Zeitung 1786, No. 146.

A record concerning the earthquake of February 27, 1786, hand-written in a Bible title page by Václav Kašlík in Valašské Meziříčí has recently been found by Jan Kozák, now in New Kozák Collection in Prague. This record well supports the IDPs for this event. Its English translation, provided by Jan Kozák, reads:

> *On Feb 27, 1786, the earthquake occurred on 4 h morning in Valašské Meziříčí, however, without any harm. During half a minute of [last] hour, rumbled as a thunderous sound, similarly as on the water [surface] wind, moving to-and-fro as a canvas craddle, an elbow-[distance]; during this moving also the dense quakes were felt, similarly as a man besieged by ague. And I, Václav Kašlík, as I felt it, I speak about is. [Additionally] it seemed to me that the [house] walls at this rocking got displaced.*

The IDP's of the 3 main earthquakes recorded in the years 1785–1786 are presented in the same scale in Fig. 1.11a–c.

There were assessed 18 IDP's for August 22, 1785 earthquake and 47 IDP's for the one on February 27, 1786. Thus, epicenter location (φ, λ) by the MEEP201 program was carried out only for threshold 3, i.e., for all IDP's without differentiation in respect to IQF. The same parameters α, k and ΔI_0 as for December 03, 1786 earthquake were assumed, i.e., $\alpha = 0.005$, isoseismal spacing $k = 3.0$, and margin of intensity increase $\Delta I_0 = 0.25$. Epicenter parameters are as follows:

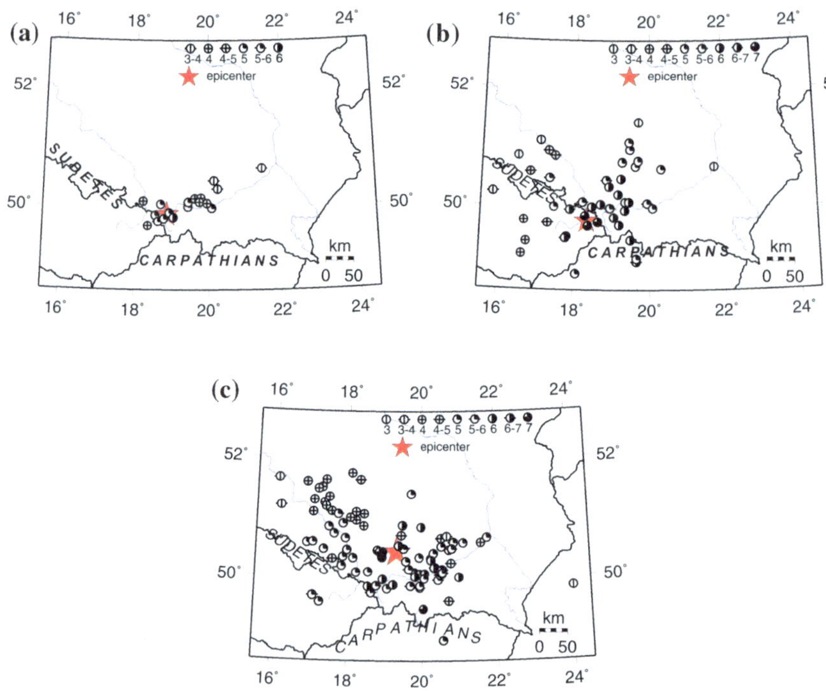

Fig. 1.11 Intensity data points (IDPs) in the EMS scale of the main earthquakes recorded in years 1785–1786. **a** August 22, 1785, **b** February 27, 1786, **c** December 3, 1786

August 22, 1785
$I_o = 6$ (EMS); $h = 16$ km; $M_M = 4.5$; $M_L = 4.6$, $\sigma = 0.37$; $M_w = 4.3$, $\sigma = 0.29$
Centroid method: $\varphi = 49.9$ N $\lambda = 18.8$E $\pm$ 16 km (assumed as the epicenter)
MEEP method: $\varphi = 49.7$ N $\lambda = 19.1$E $\pm$ 14 km
Pairwise method $\varphi = 49.4$ N $\lambda = 19.3$E $\pm$ 55 km

February 27, 1786
$I_o = 7$ (EMS); $h = 35$; $M_M = 5.4$; $M_L = 5.6$, $\sigma = 0.39$; $M_w = 5.3$, $\sigma = 0.30$
Centroid method: $\varphi = 49.8$ N $\lambda = 18.3$E $\pm$ 26 km (assumed as the epicenter)
MEEP method: $\varphi = 49.6$ N $\lambda = 18.6$E $\pm$ 14 km
Pairwise method: $\varphi = 49.7$ N $\lambda = 18.7$E $\pm$ 23 km

The results of Centroid and MEEP methods of epicenter location are in satisfactory agreement for August 22, 1785 earthquake. All the methods are in satisfactory agreement for epicenter location of February 27, 1786 earthquake. The epicenter values got by the Centroid method were assumed to be the most objective location of the epicenters. The outputs of MEEP201 program for 3 main earthquakes recorded in the years 1785–1786 are presented in Table 1.3.

1.8 Discussion of Results and Conclusions

Tectonic events in the years 1785–1786 encompassed not only the Outer Western Carpathians but also the Sudetes and probably the Inner Western Carpathians. The earthquake of February 27, 1786, on 04 pm, $I_o = 7$ (EMS), was preceded by the one that occurred about 3 hours earlier, at 01 pm, $I_o = 5$–6 (EMS), located about 50 km northwest of the main earthquake (see Sect. 1.2). Another earthquake occurred on February 13, $I_o = 5.5$ (MCS) in the Eastern Sudetes (Olczak 1966b; also see Sect. 1.2). Seismic events recorded NW of the Pieniny Klippen Belt (PKB) since the XVIII century, and those occurred in the years 1785–1786 are presented in Fig. 1.12.

Of the three main earthquakes recorded in the years 1785–1786 in the Outer Western Carpathians, the IDP's of December 3, 1786, earthquake are most scattered. This is seen in epicenter location uncertainties (Table 1.3). The dispersion of the IDP's might be a result of poor quality of macroseismic data and errors of intensity assessments, but the reason might be as well that it was not a single earthquake. Such a hypothesis, of two earthquakes, has been suggested by Réthly (1952) and Kárník et al. (1958).

One of the highest IDP's of this earthquake was recorded in Nowy Targ, at a distance of about 165–175 km from other places of $I_{max} = 7$, i.e., in Tarnowskie Góry (Q = A, q = a) and next to Tarnowskie Góry, at Świerklaniec and Rybna (both Q = B, q = d), and in Bytom (Q = B, q = d) (see Fig. 1.9). Nowy Targ has only one significant data source, i.e., the Brünner Zeitung

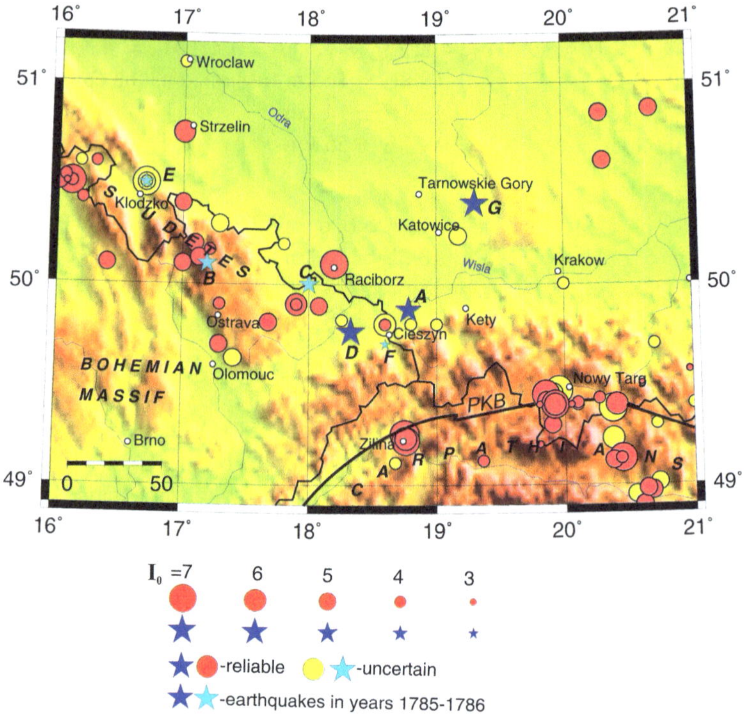

Fig. 1.12 Seismic events since the XVIII century in the Western Carpathians, NW of the Pieniny Klippen Belt (PKB), and in the Sudetes. Epicenter data of events in the Czech Republic and Slovakia after Schenková et al. (1999). Reliable: $(\Delta\phi, \Delta\lambda) < 50$ km, $\Delta I_0 \leq 1$; uncertain: $(\Delta\phi, \Delta\lambda) \geq 50$ km, $\Delta I_0 > 1$. Events: *A* – August 22, 1785; *B* – February 13, 1786; *C* – February 27, 1786, H = 01 h; *D* – February 27, 1786, *H* = 04 h; *E* – March 04, 1786; *F* – October 03, 1786; *G* – December 03, 1786

(1786), No. 103[4]; although not the witness' account, it has given a reliable summary of the felt event:

> *... An earthquake on (December) 3 was felt a few times; it affected our Nowy Targ to a greater extent than elsewhere. That day, between 5 and 6 o'clock on the afternoon, suddenly, a terrifying underground rumble could be heard for a few minutes. Right away, this was followed by two tremors, one after the other, and about 5 min later there came a third one, causing deadly fear among inhabitants....*
>
> *....It was so strong that - as was noticed later - rocks in the Carpathians cracked at a boot's width apart...*
>
> *...Here in Nowy Targ, almost in all houses, stoves collapsed. The only two stone houses cracked from the foundation to the roof so much that a man's thumb could be stuck in these fissures. The wooden houses which were poorly set in foundations were moved and twisted so that the poor people will have to work in order to put their dwellings back in order...*

---

[4] The quotations, in original language, are in the source materials: Guterch et al. (2015), this issue.

This Brünner Zeitung's note mentioning the Carpathians is the only source of information confirming that the earthquake was felt in the area near to Nowy Targ. Izydor Chrościński (1739–1811) mentioned only that the earthquake was felt in the Carpathians without any details (Chrościński 1896). Would it mean that the intensities percepted in Nowy Targ were caused by another earthquake that occurred somewhere near Nowy Targ? Could the deep tectonic process of the years 1785–1786, in the Outer Western Carpathians, trigger the seismic activity of the Orawa-Nowy Targ (ONT) Basin located 15 km south of Nowy Targ? There are no distinct data to indicate that the earthquake of December 3, 1786, was felt in the ONT Basin. The abovementioned remark in the Brünner Zeitung could have concerned this Basin, but other areas as well.The intramontane ONT Basin is located at the northernmost part of the Pieniny Klippen Belt. Seismic activity along the Pieniny Klippen Belt is recognized in many areas (Procházková et al. 1986; Kozák et al. 2002). Historical seismicity of the ONT Basin is hardly known, but contemporary seismicity has been monitored there for $I > 3$ since the last decade of the XX century (Guterch et al. 2005; Guterch 2006). The ongoing tectonic process in the ONT Basin is confirmed by seismic shocks of $M < 2.0$ that have been recorded there by a set of local seismic stations since the year 2008 (Plesiewicz and Wiszniowski, this issue). On November 30, 2004, the earthquake of $M_W = 4.4$, $I_0 = 7$, occurred in the ONT Basin, followed by a series of about 300 aftershocks recorded over nearly a year (Guterch 2006). The foci of the November 30, 2004, earthquake, as well as other known seismic events recorded there, are shallow, less than 5 km. Thus, felt intensities are strongly attenuated with distance.

It seemed possible that time observations of felt macroseismic phenomena on December 3, 1786, might help to differentiate between these hypothetical two earthquakes. There are 15 records of the time when the earthquake was felt, pointing that the earthquake occurred between 05 and 06 pm. Some data among them are more precise, i.e., on 04:45 pm (1 record), 05:10 pm (1 record), and 05:15 pm (4 records); see Fig. 1.13. Among 5 localities where the earthquake was felt with $I_{max} = 7$, the time has been recorded in Tarnowskie Góry and Nowy Targ. It was observed at 05:15 pm in Tarnowskie Góry, after Schlesische Privilegirte Zeitung (1786, Dec. 11). The time record in Nowy Targ is more complicated (Brünner Zeitung 1786, No. 103); as quoted above, it was felt repetedly a few times on December 3. Foreshocks of the December 3, 1786, event were not recognized. Thus, the phenomena observed at the beginning of the event at Nowy Targ could not be foreshocks. It seems possible that the first recorded underground rumbles could have been caused by an earthquake with its epicenter far away from Nowy Targ. Possibly, strong macroseismic phenomena recorded 5 min later could have been caused by another earthquake located much nearer, i.e., in the Inner Western Carpathians (ONT Basin?).

The data have not resolved the question whether two earthquakes occurred on December 3, 1786, but it has not excluded such a possibility. Moreover, the evidence of an aftershock on December 4, 1786 at Zator, located close to Kęty ($I = 6$ for the main event), let us suppose that it might be the area of another earthquake occurrence. The epicenters of the main earthquakes recorded on 22 August, 1785; February 27, 1786; and December 3, 1786, against the tectonic framework of Poland (after Narkiewicz et al. 2011, modified) are presented in Fig. 1.14.

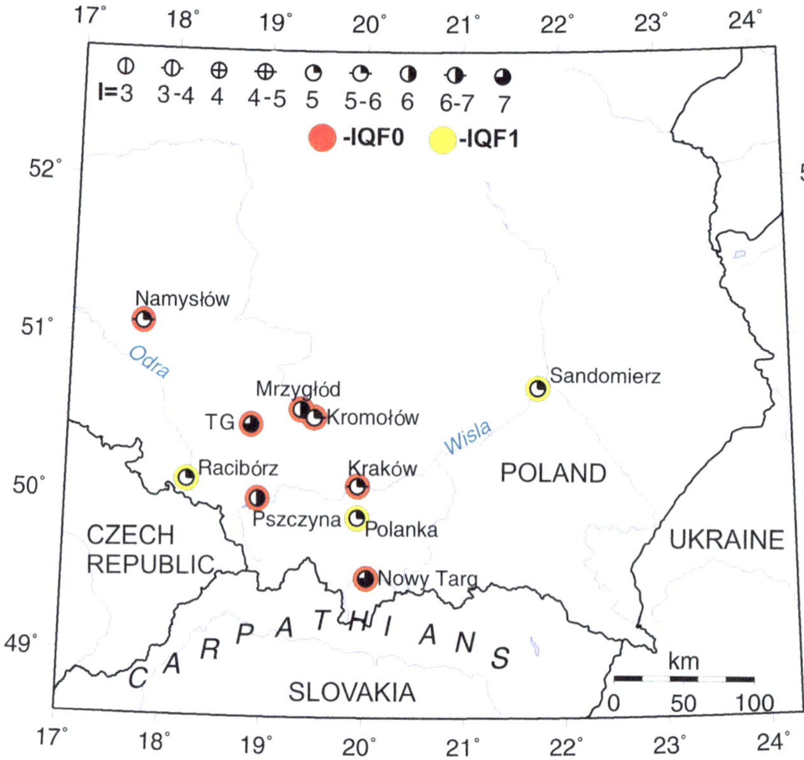

Fig. 1.13 Localities where the times of macroseismic phenomena were recorded on December 3, 1786 (IQF—intensity quality factor): Kromołów—after 05 pm, Kraków—ca 05:10 pm, Mrzygłód—05:15 pm, Namysłów—04:58 pm, Nowy Targ—after 05 pm, Polanka—05:15 pm, Pszczyna—04:45 pm, Racibórz—05:10 pm, Sandomierz—05–06 pm, Tarnowskie Góry (TG)—05:15 pm

Epicenter of the December 3, 1786 earthquake was assumed as the most objective, taking into account IDP's of the best quality available this day. This epicenter points to the Upper Silesian Block, south of the Kraków-Lubliniec Suture Zone, where a seismic event of $I_0 = 5$ was recorded in February 1837. This is a region of high induced seismicity, caused by coal mining carried out there since the XIX century. If this earthquake was not a single event, it would imply that the epicenter is a mean value. The extremely high dispersion of IDP's let us suspect that it might have been a seismic swarm. Besides the event with epicenter in the Upper Silesia, it is possible that the swarm included a second earthquake that occurred between Kęty and Cieszyn (see Guterch et al. 2015, this issue), in the epicenter area of August 22, 1785, and February 27, 1786, earthquakes; and another one in the Inner Western Carpathians (Orawa-Nowy Targ Basin?).

The main earthquakes of the years 1785–1786 are exceptional seismic events, historically known and instrumentally recorded in the Outer Western Carpathians, where Carpathians thrust over the Upper Silesian Block, in terms of

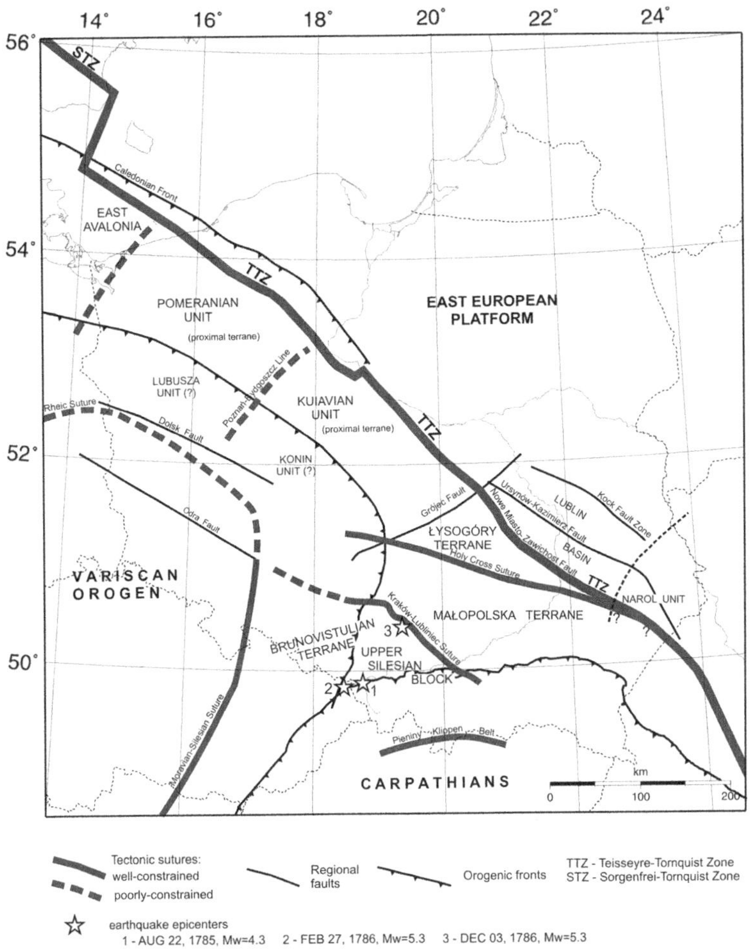

Fig. 1.14 Epicenters of the main earthquakes in the years 1785–1786 against the tectonic framework of Poland (after Narkiewicz et al. 2011, modified)

their magnitudes and depths of epicenters. In spite of problems with an epicenter assessment of December 3, 1786, earthquake, it can be stated that the main earthquakes recorded in years 1785–1786 occurred in the area of the Upper Silesian Block close to the tectonic junction of the Carpathian and Variscan orogenic fronts. This region deserves special attention in studies of seismic hazard and monitoring of local seismic micro activity should be carried out there.

Acknowledgments The inspiration for reinterpreting the December 3, 1786, earthquake and a search for new archival materials came from the late Professor S.J. Gibowicz (1933–2011), our mentor and head of the Department of Seismology at the Institute of Geophysics from 1979 through 2006.

Thanks are due to many of our colleagues from the Institute of Geophysics, Polish Academy of Sciences, for their cooperation. We are particularly grateful to Ms. Dorota Stempowska, many-year staff member of the Seismological Observatory at Książ, for exquisite translations of XVIII century German documents. Special thanks go to Edward Gaczyn´ski for his help in preparing the illustrations for publication.

We also acknowledge the contribution of Professor Jan Kozák, co-editor of this book, who provided archival materials relating to earthquakes felt in Poland in the years 1785–1786, taken from his own collections and the archives of the Geophysical Institute in Prague.

We are indebted to Dr. A. Badurová at the central Library of the Academy of Sciences in Prague for transferring the original text written by hand about the 1786 earthquake into understandale form which—after translation in English—is cited in this chapter.

Thanks are also due to Dr. Roger M.W. Musson, British Geological Survey, for making available to us the software MEEP (Macroseismic Estimation of Earthquake Parameters) version 2.0.

Special thanks to Anna Dziembowska for encouragement, assistance and help in preparing the whole monograph.

References

Alexandrowicz ST (1956) Stary obryw skalny w Tyńcu koło Krakowa. Biul Inst Geol 108:5–12 (in Polish)

Annales Ecclesiae Cracoviensis (1978) Annales Ecclesiae Cracoviensis, XII-XIV saec., Rocznik Kapituły Krakowskiej, In: Kozłowska Budkowa Z (ed) Monumenta Poloniae Historica series nova, vol V. Warszawa, year 1978, p 84 (in Latin)

Brünner Zeitung (1786) No. 99, p 787; No. 101, pp 799–800; No. 102, p 811; No. 103, pp 817–818; No. 104, pp 825–826 (Brno, Státní Oblastní Archiv, Beilage zu Brünner Zeitung, cod. I440), Brno (in German)

Carosi JF (1789) Extrait d'une lettre de M. de Carosi, Capitaine au service du Roi & de la république de Pologne, & Correspondant de l'Académie, á Mogila près de Cracovie, in: Nova Acta Academiae Scientiarum Imperialis Petropolitanae, vol V. Petersburg, pp 35–38 (in French)

Cecić I, Musson RMW, Stucchi M (1996) Do seismologists agree upon epicenter determination from macroseismic data? A survey of ESC working group "macroseismology". Ann Geofis 39:1013–1027

Chrościński I (1896) Izydora Chrościńskiego Kronika z lat 1784–1794 z przydaniem aktów fundacyjnych kolonii akademickiej w Tarnowie z lat 1559 i 1578, jako też Raportu o jej stanie z roku 1766, J. Leniek, (ed) Kraków, p 79 (in Latin)

Cieszyn AP, manuscript (1767–1822), Achiwum Państwowe w Cieszynie, V Abtheilung des Gedenk-Bücher der Stadt Teschen vom Jahre 1767 bis 1822, fasc. 74, fol. 114 (in German)

Długosz J (1873, 1877) Joannis Dlugossii seu Longini Canonici Cracoviensis Historiae Polonicae Libri XII. In: A. Przezdziecki (ed.) Joannis Długosz Senioris Canonici Cracoviensis Opera Omnia, Kraków, Cracoviae, ex Typographiae Kirchmayerianea XI (1873), XIII (1877) (in Latin)

Draber D, Guterch B, Lewandowska-Marciniak H (2000) Local earthquakes recorded by the Polish seismological stations 1993–1994. Publs Inst Geophys Pol Acad Sci B-22(331):3–7

Eisinger U, Gutdeutsch R, Hammerl Ch (1992) Historical earthquake research—an example of interdisciplinary cooperation between geophysicists and historians. In: Gutdeutsch R, Grünthal G, Musson R (eds) Historical earthquakes in Central Europe, vol 1. Vienna, pp 33–50

Encyklopedia Krakowa (2000) Stachowski AH (ed) PWN, Warszawa, Kraków, p 561 (in Polish)

Frankel A (1994) Implications of felt area-magnitude relations for earthquake scaling and the average frequency of perceptible ground motion. Bull Seismol Soc Am 84:462–465

Gasperini P, Bernardini F, Valensisei G, Boschi E (1999) Defining seismogenic sources from historical earthquakes felt reports. Bull Seismol Soc Am 89:94–110

Gazeta Warszawska (1786) Dec. 09, No. 98-supplement; Dec. 13, No. 99; Dec. 16, No. 100; Dec. 23, No. 102; Dec. 30, No. 104 Warszawa (in Polish)

Gazeta Wileńska (1786) Dec. 16, No. 50-supplement; Dec. 23, No. 51; Dec. 30, No. 52, Wilno (in Polish)

Grabowski A (1900) Kraków i jego okolice /opisał historycznie Ambroży Grabowski; Friedlein DE (ed) Kraków, pp 187–188 (in Polish)

Grünthal G (1992) The central German earthquake of March 6, 1872. In: Gutdeutsch R, Grünthal G, Musson R (eds) Historical earthquakes in central Europe, vol. 1. Wien, pp 51–69

Grünthal G, Walström R, Stromeyer D (2009) The unified catalogue of earthquakes in central, northern and northwestern Europe (CENEC)—updated and expanded to the last millennium. J Seismol 13:517–541

Gutdeutsch R, Hammerl Ch, Mayer I, Vocelka K (1987) Erdbeben als historisches Ereignis, Die Rekonstruktion des Bebens von 1590 In: Niederösterreich, Springer, Berlin, Heidelberg, 222

Guterch A, Grad M, Keller GR (2007) Crust and lithospheric structure—long range controlled source seismic experiments in Europe. In: Schubert G, Romanowicz B, Dziewonski A (eds) Treatise on geophysics, vol 1. Elsevier, Amsterdam, pp 533–558

Guterch B, Lewandowska-Marciniak H (2002) Seismicity and seismic hazard in Poland. Folia Quater 73:85–99

Guterch B, Lewandowska-Marciniak H, Niewiadomski J (2005) Earthquakes recorded in Poland along the Pieniny Klippen Belt, Western Carpathians. Acta Geophys Pol 53:24–45

Guterch B (2006) Seismic events in the Orawa-Nowy Targ Basin, Western Carpathians, November 30, 2004–December 2005, Acta Geodyn Geomater 3(143):85–95

Guterch B (2009) Sejsmiczność Polski w świetle danych historycznych (Seismicity in Poland in the light of historical records). Przegląd Geologiczny 57(6):513–520 (in Polish)

Guterch B, Lewandowska-Marciniak H, Mazur M (2015) Source Materials to the Outer Western Carpathian Earthquake of December 3, 1786, this issue

Hoff KEA (1841) Chronik der Erdbeben und Vulkanausbrüche, Gotha, 78 (in German)

Jeitteles HL (1860) Versuch einer Geschichte der Erdbeben in den Karpathen und Sudeten Landern bis zu Ende des achtzehnten Jahrhunderts, Zeitschrift d. Deutsch Geol Ges 12:287–349 (in German)

Kárník V, Michal E, Molnár A (1958) Erdbebenkatalog der Tschechoslowakei bis zum Jahre 1956. Geofysikalni Sbornik 1957, Travaux de l'Institut Geophysique de l'Academie Tchecoslovaque des Sciences, No. 69, pp 411–598 (in German)

Kárník V (1961) Die Seismizität der Westkarpathen, Geofysikální Sborník 1960, Travaux de l'Institut Geophysique de l'Academie Tchecoslovaque des Sciences, No. 134, pp 265–283

Kárník V (1968) Seismicity of the European area, part 1. Academia, Praha

Kárník V, Procházková D, Schenková Z (1977) Earthquake epicentres and macroseismic intensity in the Carpatho-Balkan region. Acta Geologica Academiae Scientiarum Hungaricae, Tomus 21(4):309–324

Kęty MO manuscript, (1785–1795) Muzeum Okręgowe, Oddział w Kętach, (Regional Museum, Branch in Kęty) Księga Cechu Tkackiego Barchanickiego, fasc. H/697 (in Polish)

Koláček F (1916) Zemětřesení na Moravě a v rakouském Slezsku, 23, Sborník České Společnosti, Praha, pp 153–160 (in Czech)

Kozák J, Buben J, Vencovský M (2002) Delineation and segmentation of the Peripierninian lineament. Acta Geophys Pol 50(2):179–196

Kraków AKAP-2, manuscript (1786) Archiwum Prowincji Zakonu O.O. Kapucynów w Krakowie, Kronika Klasztoru Krakowskiego, fasc. AKK. 95, fol. 240 (in Latin)

Kraków AKK, manuscript (1786–1787) Archiwum Kapituły Katedralnej w Krakowie, Acta Actorum Capituli 1776–1787, fasc. 25, fol. 714–715 (in Latin)

Kraków AKMK-1, manuscript (1786) Archiwum Kurii Metropolitalnej w Krakowie, fasc. A. Ep. 112, List Pasterski Prymasa Polski, fol. 203 (in Polish)

Kraków AKMK-2, manuscript (1786) Archiwum Kurii Metropolitalnej w Krakowie, fasc. A. Ep. 112, Zalecenie do kościołów farnych i zakonnych, fol. 252–253 (in Polish)

Kraków AKMK-6, manuscript (1786) Archiwum Kurii Metropolitalnej w Krakowie, fasc. Ag. 12, Raport Miesiąca Grudnia z Dekanatu Kijskiego Roku 1786, fol. 391 (in Polish)
Kraków AP-1, manuscript (1786) Archiwum Państwowe w Krakowie (State Archives in Kraków), Archiwum O.O. Augustianów, fasc. 22, Kronika Klasztoru, fol. 88, fol. 109 (in Latin)
Krzyżanowski T (1900) Wspomnienia mieszczanina krakowskiego z lat 1768–1807, Prokesch W (ed) Kraków, 29–30 (in Polish)
Laska V (1902) Die Erdbeben Polens, Mittheilungen der Erdbeben Commission, Keis. Ak. d. Wiss. in Wien, Wien, No. 8, pp 1–36 (in German)
Leonhard R, Voltz W (1896) Das mittelschlesiche Erdbeben vom 11. Juni 1895, Zeitschrift des Ges. f. Erdkunde zu Berlin, Bd. 31, p 18 (in German)
Lwowskie Tygodniowe Wiadomości, 1786, Dec. 20, No. 94; Dec. 23, No. 95; Dec. 27, No. 96; 1787, Jan. 03, No. 98, Lwów (in Polish)
Mallet R, Mallet JW (1858) The earthquake catalogue, London, p 15
Miechowita M (1986) Maciej z Miechowa, Chronica Polonorum Mathiae de Mechovia, Liber Tercius, editor: H. Wietor (1521); facsimile, 1986, Sł. Tabkowski (ed.), Krajowa Agencja Wydawnicza, Kraków p. 157, p. 159 (in Latin)
Musson RMW, Jiménez M-J (2008) Macroseismic estimation of earthquake parameters, NERIES project report, Module NA4, Deliverable D3
Musson RMW (2009) MEEP 2.0 User guide, British geological survey, earth hazards and system programme, open report R/09/045
Narkiewicz M, Grad M, Guterch A, Janik T (2011) Crustal seismic velocity structure of southern Poland: preserved memory of a pre-Devonian terrain accretion at the East European Platform margin. Geol Mag 148(2):191–210
Olczak T (1965) Trzęsienie ziemi z dnia 22 VIII 1785 roku w Beskidzie Śląskim (Earthquake of 22 August 1785 in Silesian Beskids), Przegląd Geologiczny X (XVIII)(3–4):227–237 (in Polish)
Olczak T (1966a) Sur la region epicentrale et la profondeur d'un foyer seismique du 3 XII 1786 dans les Carpathes occidentales polonaises. Acta Geophys Pol 14(1):61–67 (in French)
Olczak T (1966b) O Karpackim trzęsieniu ziemi z dnia 27 II 1786 roku i towarzyszących mu ruchach sejsmicznych na przedpolu Sudetów Wschodnich (On Carpathian earthquake of 27 February 1786 and accompanying seismic motions in the East Sudetic forefield) Przegląd Geologiczny, XI (XIX), No. 1, pp 3–21 (in Polish)
Pagaczewski J (1972) Catalogue of earthquakes in Poland in 1000–1970 years. Publs Inst Geophys Pol Acad Sci 51:3–36
Pamiętnik Historyczno Polityczny (1787) Pamiętnik historyczno-polityczny przypadków, ustaw, osób, miejsc i pism wiek nasz szczególniey interesujących, 1786; P. Świtkowski (ed) Warszawa, vol IV, pp 237–238 (in Polish)
Perrey A (1846) Memoire sur les tremblements de terre dans le bassin du Danube, Lyon, p 361 (in French)
Pohludka S (1897) Erdbeben in Mahren und Schlesien, Mitteilungen des Naturwiss. Vereines in Tropau, No. 5, III. Vereinsjahr, p 95 (in German)
Procházková D, Dudek A (1980) Some parameters of earthquake originated in central and eastern Europe. Travaux Geophysiques XXVIII(538):43–82
Procházková D, Dudek A, Misař Z, Zeman J (1986) Earthquakes in Europe and their relation to basement structures and fault tectonics, Rozpravy ČSAV, Rada Matem. A Přírodních Věd Praha 96(2):77
Procházková D, Kárník V (eds) (1978) Atlas of isoseismal maps, central and Eastern Europe. Geophysical Institute of the Czechoslovak Academy of Sciences, Prague, pp 4–6
Remeš M (1930) Zemětřesení pozorovaná na Moravě a ve Slezsku, Zvlastni otisk z Věstníku Klubu Přírodovědeckého v Prostějově za rok 1928–1929, Prostějov, pp 1–10 (in Czech)
Réthly A (1952) A Kárpátmedencsék Földrengésci 455–1918. Akademiai Kiado, Budapest (in Hungarian)
Rocznik Krasińskich (1878) In: Bielowski A (ed) Monumenta Poloniae Historica, vol III. Lwów, p 132 (in Latin)

Rocznik Magistratu Wrocławskiego (1878) In: Bielowski A (ed) Annales Wratislaviensis, Monumenta Poloniae Historica vol III. Lwów, p 682 (in Latin)
Rocznik Małopolski (1878a) Annales Poloniae Minoris, kod. Kuropatnickiego (XV–XVI saec.). In: Bielowski A (ed) Monumenta Poloniae Historica vol III. Lwów, p 168 (in Latin)
Rocznik Małopolski (1878b) Annales Poloniae Minoris, kod. Szamotulski (1471). In: Bielowski A (ed) Monumenta Poloniae Historica, vol III. Lwów, p 169 (in Latin)
Rocznik Wrocławski Dawny (1878) In: Bielowski A (ed) Annales Wratislaviensis, Monumenta Poloniae Historica, vol III. Lwów, p 682 (in Latin)
Schenková Z, Schenk V, Kottnauer P, Guterch B, Labák P (1999) Earthquake catalogue for the Czech Republic, Poland and Slovakia (manuscript)
Schlesische Privilegirte Zeitung (1786) Dec. 09, No. 145; Dec. 11, No. 146; Dec. 20, No. 150; Dec. 23, No. 151, Wrocław (in German)
Schlesische Provinzialblätter (1786), vol 4, No. 12, pp 593–596, Oleśnica (in German)
Schmidt J (1858) Untersuchungen über das Erdbeben am 15 Januar 1858, Wien, p 68
Sieberg A (1940) Beitrage zum Erdbebenkatalog Deutschlands und angrenzender Gebiete für die Jahre 58 bis 1799, Mitteilungen des Deutschen Reichs-Erdbebendienstes, Jena, No. 2 (in German)
Słownik Geograficzny (1892) Słownik Geograficzny Królestwa Polskiego i Innych Krajów Słowiańskich, Bronisław Chlebowski (ed) Tom XII, p 714 (in Polish)
Sponheuer W (1960) Methoden zur Herdtiefenbestimmung in der Makroseismik, Freiberger Forschungshefte, Geophysik, C88, 117p (in German)
Stryjkowski M (1980) Kronika polska, litewska, żmódzka i wszystkiej Rusi Macieja Stryjkowskiego, Wydawnictwa Artystyczne i Filmowe, Warszawa, p 293 (in Polish)
Wojciechowski B (1895) Fragment z dziejów szkół kaliskich, in: Archiwum do dziejów literatury i oświaty w Polsce, vol VIII. Kraków, pp 323–324 (in Polish)

Chapter 2
First Macroseismic Maps in Southern Poland in the Late 19th Century

Jan Kozák

Abstract The principles of macroseismic earthquake observation and classification in pre-instrumental era are briefly outlined and commented, which—in the course of the 19th century—resulted in making macroseismic intensity maps, forerunners of present maps of seismicity. The achievements in macroseismic earthquake study found their reflections in Central Europe, namely in Poland: for the second half of the 19th century (1858–1901) a total of 19 macroseismic maps were found for three Sudetic events and two Polish border area earthquakes in Saxony and NW Slovakia. They are reproduced and discussed below.

Keywords Macroseismic earthquake treatment · Macroseismic maps · Macroseismic intensity scale · South Poland historical earthquakes

2.1 Geopolitical Note

In the 19th century, the term *Poland*, and its territorial content, was variable in shape and size due to complex geopolitical development of Polish state in the past centuries. Similarly, the area of Silesia (which was more-or-less stable in shape and size), has varied considerably as concerns its geopolitical status, its population's alienisms and ethnic profile.

Since the cogitations given below deal with the historical seismicity of the region in question, one could be confused by the above-mentioned diversity. To avoid it we—in the following text—shall understand Poland, with its southwestern part, Silesia, exclusively as it has been settled and confirmed in 1945.

J. Kozák (✉)
Institute of Geophysics, Academy of Sciences of the Czech Republic,
Prague, Czech Republic
e-mail: kozak@ig.cas.cz

B. Guterch and J. Kozák (eds.), *Studies of Historical Earthquakes in Southern Poland*, GeoPlanet: Earth and Planetary Sciences,
DOI 10.1007/978-3-319-15446-6_2

2.2 Macroseismic Cartography in Transalpine Europe in the Second Half of the 19th Century

2.2.1 Introductory Remarks

Since the early Modern Age, European naturalists and philosophers strengthened their effort to better understand the perplexing phenomenon of *earthquakes* initiated deep in the Earth during which, without any warning, enormous portion of mechanical energy was suddenly released, demolishing settlements and killing people. In the late 18th century it became obvious that without a deeper insight into the complex behavior of the inner Earth processes, geosciences could not proceed to reply the main question: when these enigmatic forces will strike again, i.e., to the forecast an earthquake approach (this aim—by the way—remains unsolved up to the present time for the great deal of seismically active zones).

The first serious studies on earthquakes were presented by Swiss and English savants at the second half of the 18th century, in the time when the 1755 Great Lisbon Earthquake was subjected to a detailed investigation by numerous intellectuals of that-time Europe; for details see, e.g., Davison (1978, 2nd ed.), Reid (1914), Kendrick (1956), Schmidt (1980) and others. Ch. Davison in his book describes the main ideas on an earthquake presented by John Bevis, who lived in 1693–1771, as well as Bertrand (1757) and Michell (1761). After that glorious period of the early seismology—after Ch. Davison—a long pause occurred, from the 1760s to ca 1847, characterized by a kind of retardation of the progress in earthquake studies.

Of course, the *pause* seen by Ch. Davison was not quite tangible as concerns achievements in seismology. It was especially the long series of strong Calabrian earthquakes from 1783 to 1791 which excited Europeans similarly as the previous Great Lisbon earthquake of 1755; in Calabria six consequent strong earthquakes caused medium or large damage to the buildings within the area of more than 100 000 miles^2, which corresponds to ca ¼ million km^2. The vast regions of Calabria and eastern Sicily, which were hit by these earthquakes, were carefully studied by Italian naturalists, who summarized macro-seismic observations in their reports and books, see, e.g., Vivenzio (1783 and 1788), Dolomieu (1785) and others; confer also Barbano et al. (1980) and Placanica (1985).

As for the state of geosciences in the mid-19th century, many a concrete achievement obtained earlier in the field of investigation of seismic energy awakening, propagation and interaction with the Earth surface was forgotten. On the other hand, the naturalists succeeded in learning more clearly *which way* they should follow in the next research in order to proceed in the earthquake study; it was found and affirmed at the break of the 18th century that especially the seismic data of macroseismic nature detected by human observers on the Earth surface—if properly treated and interpreted—may shed light upon the invisible underground processes of seismic energy release.

This new insight into the earthquake study in Europe appeared at the break of the 18th century as being accompanied by a new phenomenon: by the expansion of savants' interest in the field investigation of seismic events *from the Mediterranean zones towards the north*, to the transalpine regions. The first steps in this process were done for the western parts of the continent, later for Central Europe, up to Poland and Hungary. This *transalpine* progress, which became clearly evident since the first years of the 19th century, and which strengthened later in that century, did not occur by chance: in that time German population went on a mass-moving eastward from their traditional seats around Rhine, Danube and Elbe Rivers so that in the course of 18th–19th centuries a large part of Central Europe appeared under German influence, expressed in political and cultural Germanic supervision. As concerns the natural sciences development in this region, one has to admit that the strong German states, Prussian, Hapsburgs' and Bavarian, undoubtedly appeared in a positive light in numerous aspects. Let us look—first of all—at the situation in geosciences, namely in early macro-seismology, which gradually developed as an important new discipline of geophysics in transalpine Europe, in which also Poland was involved.

Local transalpine geo-savants, not yet having the desired *Erdbebenmesser* (=device to measure an earthquake) for collecting and treating the *objective* seismic data at their disposal, developed instead a practical system of *subjective* macroseismic data collection, selecting and valuing by (more-or-less instructed and trained) human observers. These observers recorded visible seismic damage and classified it by means of answering the questions given in advance prepared *earthquake questionnaires*; possible variations in classification of the same damage by individual observers were compensated by averaging a large number of observations. The data obtained in such a way enabled to be sized according to the level of seismic damage advised by pertinent macroseismic intensity scale(s).

Soon in the 19th century it became clear that the "strength" of the earthquake is reflected—besides the observed degree of seismic damage on the Earth's surface—also by the *size of the stricken territory*. This fact led directly towards constructing macroseismic maps, alias geographical maps, in which the observed macroseismic data were localized according to their geographical coordinates λ, φ.

Later in the 19th century it appeared reasonable to investigate earthquakes not as separate events and classify them according to arbitrary criteria, but to apply an unified approach to earthquake studies as to mutually dependent phenomena and to analyze them using the same methodical tools, such as, e.g., quantity of seismic damage (=macroseismic degrees of damage), using the unified scale of damage, etc. Such a general unification appeared as an important condition for comparative studies of individual seismic territorial, time and intensity situations.

Having in mind this general effort aimed on an earthquake study, which can be considered as a rational and integral whole, we may classify the 19th century as a period of macroseismic pre-instrumental seismology birth. This period was at its end crowned by the first macroseismic maps and maps of seismicity on one hand, and by the first seismometers by J. Ewing and D. Milne at the turn of the 1870s, on the other hand. Let us now characterize briefly the individual steps in this macroseismic study advancement during the 19th century.

2.2.2 *Quantity of Earthquake "Strength"; Earthquake Intensity and Intensity Scales*

The seismic effects occurring on the Earth's surface due to seismogenic displacements in the Earth interior, as seen by human observers, were expressed by degrees of macroseismic intensity I (symbol I_o being reserved for maximum I value in the earthquake epicenter zone). It follows that the I-values were derived by means of statistics of macroseismic observations of humans, on animals, and man-made structures and natural formations.

Quite naturally, due to diversity of the Earth surface and local seismic activity of a given region, and also due to space/time varying culture and traditions of the residing local population, also formulation of I-degrees and consequently of macroseismic intensity scales *varied considerably region to region*. This diversity, reflected by the variable content of individual I-degrees, and also by total number of damage degrees of a given scale, often hampered the desirable general comparison of individual world earthquakes.

In the first ancient and medieval reports on the earthquakes, the described events were, in most cases, named according to the principal town in the affected area; for strong earthquakes the whole region (province, county) was named. Later, in the 17th and 18th centuries, for some Italian earthquakes, the seismic damage to buildings was differentiated in the pertinent maps, e.g., 4 levels of damage for the 1627 Puglia event (Foglia 1627) or 3 levels of damage for the 1783 Calabria earthquakes (Vivenzio 1783). In the first half of the 19th century, new macroseismic scales were proposed (6-degree scale by P.N.C. Egen for the 1828 upper Rhine earthquake (Egen 1828) and by J.J. Noggerath for the 1846 mid-Rhine earthquake (Noggerath 1847)); sometimes, the names of the settlements damaged in given degree were underlined in the standard geographic maps by color corresponding to the degree of damage given in the map legend. The isoseismal lines were not drawn in most cases; however, in some earthquake maps of this period the line showing limit of (human) perception of the Earth shaking was drawn; actually, the first such map presenting this limit was prepared and presented by two Hungarian naturalists, A. Tomtsanyi and P. Kitaibel, in their report on the 1810 Mór earthquake, see below.

In the second half of the 19th century, macroseismic cartography advanced rapidly, especially due to conscientiously performed macroseismic analysis of the 1855 Visp earthquake (Wallis, Switzerland), which was presented by Volger (1856 and 1857–1858); 5 years after O. Volger's publishing of his voluminous monograph, R. Mallet presented his 3-volume work on the 1857 Great Neapolitan earthquake (Mallet 1863). Both monographs represented model examples of the high level macroseismic analysis of a strong earthquake, including cartographic expression of pertinent seismic effects.

O. Volger, German naturalist teaching mineralogy at the high school in Bern, Switzerland, studied phenomenon of an earthquake as his hobby. After the 1855 Visp earthquake he spent several years collecting macroseismic data in

Switzerland and in the adjacent parts of France, Italy and Germany, which were also affected by the Visp earthquake. Then he defined his 6-degree macroseismic scale of seismic damage (in degrees of intensity I) and plotted his data into standard geographical map of the studied region. By plotting envelopc curves for individual intensity degrees he got a standard macroseismic intensity map with the isoseismal lines as we understand it at present. Additionally, he improved his map in three points.

First, he added into his isoseismal map also thre perception limits of three weaker aftershocks, which followed the main shock.

Second, he strictly kept the curvature of his isoseismal lines according to the collected seismic macrodata plotted in his map (in contrast to other naturalists, who smoothed out "by sight" the shape of isoseismal lines, which they transformed into nice ovals, see, e.g., Mallet (1863) or Mouchketow (1903). It seems that O. Volger understood that the medium in which seismic waves propagate is not homogeneous and that due to complex geometrical texture of the medium the geometry of propagation of seismic waves may be influenced.

Third, in the right bottom corner of O. Volger's map the epicentral zone's sketch blown up is inserted to show the details of the pleistoseismal zone complexity.

We could also appreciate that he tried to mount the 1855 event into the pattern of seismically active arc of the central and western Alps.

The only weak point in O. Volger's macroseismic map can be seen in the numerical order of individual intensity degrees: namely, he denoted the strongest ($I_o = I_{max}$) intensity of the Visp earthquake as the zone **zero**, the next (still strong) zone as the zone No. **one**, then zone No. **two**, etc. up to weakest zone No. **six**. This order of intensity degrees was perhaps acceptable for one earthquake (=Visp), but it did not allow an intensity comparison with other, stronger earthquakes (unless one could agree to describe stronger events by negative numbers: −1, −2, −3, …). However, even with this small reservation, O. Volger's macroseismic map must be highly appreciated and valued.

After O. Volger's analysis published in 1857–58 it became commonly understandable how to create the macroseismic scale of intensity I, how to order its degrees corresponding to seismic damage, and how to express correctly the individual I-levels by $I_0 = I_{max}, I_{max-1}, I_{max-2}, I_{max-3}, \ldots I_1 = I_{min}$. The first macroseismic intensity scale which was internationally accepted as a standard 10- (later12)-degrees of macroseismic intensity classification was the Forel (1881) scale, in which the value I_0 always corresponded to maximum (=epicenter) intensity, while I_{min} represented the lowest level of seismic damage observed, or, in some cases, the limit of Earth trembling perception felt by a man. Later on, the M. Rossi scale or M. Rossi and F. Forel macroseismic intensity scales were repeatedly changed, complemented, their degrees being refined and formulated more in detail (Rossi and Forel 1881). The last present variant of the macroseismic intensity scale represents simultaneously the widely used *European Intensity Scale-98* (EIS 1998), which was edited and composed by G. Grünthal and published just at the turn of the 20th century (Grünthal 1998). It should be noted that the scale EIS-98 is not used all over the world as the only mandatory macroseismic intensity degree scale; in many countries of Asia and America, older scales are used as well.

As for the territory of Poland, the name and life-work of the outstanding Polish geo-savant and geosciences promoter M.P. Rudzki, professor at the Jagielonian University, astronomer, geophysicist and seismologist, who organized foundation of the first seismological observatory at Kraków in 1903, must be recalled in this context. However, regardless of great M.P. Rudzki's merits as concerns Polish seismology, he did not engage himself in macroseismic cartography; also his main seismological activities fall into the 20th century,[1] which is not considered in the macroseismic studies presented here.

Today we know that the degree of seismic damage alone does not provide the proper degree of earthquake "strength" unless the geometry of the source and seismogenic displacement and especially the depth of seismic foci are considered. (As for the earthquake foci depth, in the early times geo-savants had no reliable tool in their hand to determine it. However, in some cases they spoke about a *shallow* earthquake in the situations in which seismic effects became detectable on the Earth's surface relief.)

2.2.3 Macroseismic Questionnaires and Seismic Historical Catalogues

Since the macroseismic analysis of an earthquake is based on treatment of observational information, its proper collection, recording, and delivery are of a crucial importance for obtaining undistorted results. In principle, these observational data are—first of all—used for two reasons. First, local or higher-level administrative institutions use them to organize the rescue and aid operations to mitigate the earthquake damage and, secondly, to study the observed data carrying new information on the seismic phenomenon itself; they helped to understand better the destructive seismic process and, in an optimum case, they were believed to be considered in the terms of the desired local seismic prognosis.

Seen from this viewpoint, the collected macroseismic data (which come from unequally trained and educated humans) should be still obtained in a form characterized by maximum objectivity and minimum of subjective and emotional viewpoints so that they would allow a correct determination of the given degree of macroseismic intensity for localities visited by the earthquake in question. It has appeared in the course of the last two or three centuries that (in the countries inhabited by literate population) for the above-mentioned requirements the *macroseismic questionnaires* serve as an important tool.

The questions asked for in these question-forms can be formulated according to the macroseismic scale used so that assignments of individual *I*-values

[1] His pioneering book "Physics of the Earth" was published in Polish in 1909 and then in German in 1911 (Rudzki 1909, 1911).

to individual situations described in the reports are quick and relatively unambiguous. The fact that the questionnaires can be submitted to a large number of respondents reduces the weight of possible extreme and/or incorrect data and figures.

A questionnaire after-earthquake circularization was reported, e.g., in the Iberian Peninsula in 1755/56, after the Great Lisbon earthquake occurrence, which devastated a large part of SW Europe on Nov 01, 1755. In the last two centuries, the seismic questionnaires represented an important source of macroseismic data: today modern seismographs supply directly and in an objective form a great portion of the required information, which earlier had to be assessed by making use of macroseismic observations. In some specific cases, however, both approaches, instrumental and macroseismic, have been successfully combined up to the present.

Other important tools for seismological research are represented by seismic historical catalogues.

For the first attempt to create a historical inventory of world earthquakes, which appeared in the seismically active regions, primordially in Italy, see, e.g., Münster (1544→1618) and Rasch (1582); however, their lists contain only a limited number of earthquakes, often wrongly dated and/or located and described. The earthquake information was often many-times taken over and distorted. On the other hand, the information about *old-time* seismic events has always been valuable, because the recurrence time of strong earthquakes could be much longer than that of a human lifespan. Therefore, each true point in the long-term diagram of given zone's seismic activity is much welcome. However, it does not change the fact that all the catalogues are in a sense incomplete, which must be always taken into account.

It can be concluded that the work on the "most complete version" of the catalogues is an endless work. First, nobody can get more-or-less complete catalogue but for the recent period only; beyond that, the new, modern seismic research brings new requirements to furnish the modern catalogue records by new parameters, which cannot be detected for the historic events. Finally, modern and newly collected catalogues may seem to be relatively perfect when they are completed; however, in a couple of years they appear out of date. Yet, regardless of these weak points of catalogues, geo researchers cannot but work with them since nothing better is at their disposal.

2.2.4 Seismic Catalogues of Polish Territory and Polish Border Countries

In the countries of low seismicity, such as Poland, the first catalogues of the whole country appeared much later than in seismically active countries, such as, e.g., Italy, because seismic risk and seismic endangerment of the population were low. The first catalogue of Polish territory was published by Láska (1902), where a brief verbal description of 70 earthquakes in the time interval 768–1900 is given. Seventy years later, a catalogue of Poland for the time span 1000–1970 was published by Pagaczewski (1972); his catalogue is accompanied by a modern map of seismicity of the regions in question; for the focal zones of Poland, e.g., for Lower and Upper Silesia, where the seismicity is relatively high, local historical catalogues appeared earlier.

Seismicity of Western Carpathians was investigated by Kárník (1960).

Historical earthquakes in the territory of Czechoslovakia can be, for the period 460–1956, found in the Erdbeben Katalog der Tschechoslowakei by Kárník et al. (1957).

The recent catalogue related to the Czech and Polish Sudetes zone was compiled by Schenk and Schenková (2007).

The newest catalogue of Germany partly covering also Silesia was composed by Leydecker (2011).

Obviously, there were other old local catalogues published in the last centuries for the territory of Silesia, Saxony, Bohemia, Moravia and West Carpathians, in which some earthquakes occurring in Polish border regions are listed, similarly as some events having epicenters out of Poland, which were perceptible and/or recorded in Polish territory as well. Since the earthquakes listed in these old, mostly local catalogues were altogether accounted for in modern catalogues mentioned above, they will not be further considered in this contribution.

2.2.5 Forming Seismic Research in a Global Dimension

Under activities named in the title of this section we understand the establishment of specific seismological institutions, seismic observatories and university divisions, organizing international seismological meetings, founding specific seismic journals, etc.

In the course of the 19th century a new approach towards earthquakes became apparent in a growing rate. It was based on the global character of strong phenomena of dynamical Earth's manifestation—including earthquakes—in contrast to the early period of seismic studies in the end of the Middle Ages up to late 19th century, in which seismic studies tried to explain the mysterious forces ruling the inside of the Earth as solitaire, mutually not related categories. This tendency was reflected in a number of relevant changes. First, seismic studies left off the status of a "poor relation" struck on the famous, already well-known disciplines of geology, geography, volcanic sciences or physics. On the contrary, seismology alone became new, independent geophysical branch of geo-research lectured in newly founded university seismological departments (=cathedras) and driven in newly established seismological institutions and observatories. The achievements reached in these institutions were presented and discussed in newly organized international meetings and published in newly founded geophysical journals.

For all of the new seismological institutions, we may give the names of sites where they were established in Transalpine Europe at the second half of the 19th century: Strasbourg, Goettingen, Berlin, Królewiec/Königsberg, Muenchen, Vienna, Leoben, Zagreb, Budapest, Kraków,[2] Lwów, Czerniowce, etc. Also some earlier founded institutions (e.g., the Bergakademies in Banská Štiavnica, in Freiberg/Sachsen, Clausthal-Zellerfeld, Kraków and Leoben and national Academies of Sciences in Berlin, Vienna...) expanded their research program to seismic studies.

In Germany, geophysics got an international recognition in 1898, when E. Wiechert was appointed at the University of Göttingen as the first Professor of Geophysics. In the same year he founded the first Institute of Geophysics at this University. In 1895 E. Wiechert also organized "The 1st International conference on seismology"; this meeting was held in Strasbourg on April 1–3, 1901 (Schröder 2000; Kozák 2001).

As for new seismological journals let us mention at least Garland's Beiträge zur Geophysik (Leipzig, since 1887, regularly since 1898), Mittheilungen der Erdbeben Kommission (Vienna, since 1897) or A. Belar's Erdbebenwarte (Zagreb, since 1902).

Today, with the benefit of hindsight we may proclaim the second half of 19th century as the period of birth of scientific seismology, which made the first steps in seismic wave recording and seismogram evaluation.

Now, let us turn our attention to the southern territory of Poland, i.e., Lower and Upper Silesia and Galicia (or Sudetes and West Carpathians) to examine up to which degree the advancement of the European macro-seismology, as outlined above, was reflected in that time of occurrence of macroseismic intensity maps of earthquakes prepared in southern Poland.

2.3 The First Macroseismic Maps (MSM) of the Polish Territory

Three macroseismic maps related to the earthquakes in the transalpine region in 1810–1846 ***preceding the Polish maps*** should be mentioned, namely the maps of the 1810 Mór earthquake (in Hungary), the 1828 event (in the Lower Rhine region) and the 1846 earthquake (in the Central Rhine region).

The earthquake of 1810, Jan 14, by Mór (ca 70 km west of Budapest), damaged buildings in several villages in the broad terrain-gap in the Bakónyi Hills ridge. Earthquake intensity reached $I_0 \approx$ 8–9 MSK, perception area was equal to ca 32 000 km^2. Special commission of geo-savants was appointed at the Buda University to examine the earthquake and delegated to visit the afflicted area. Two members of the commission, P. Kitaibel and A. Tomtsányi, accompanied their report on the earthquake by a simple macroseismic map, in which the seismically damaged settlements were marked and circumscribed by a dotted line in the

[2] The first in the world geophysical chair, named Chair of Mathematical Geophysics and Meteorology, was created for the previously mentioned M.P. Rudzki at the Jagielonian University in Kraków in 1896.

map, roughly equal to the damage limit line; no intensity scale was considered (Kitaibel and Tomtsanyi 1814). By present viewpoints, the Mór earthquake map cannot compete with the standard present macroseismic intensity maps. However, the effort of the Hungarian geo-savants to express the earthquake effects in a map of the affected region must be highly appreciated: they were the first ones who re-discovered the far forgotten ideas of Italian naturalists in the 17th century to present the earthquake effects in a map form. Detailed information on the Mór earthquake and Mór macroseismic map can be found in several younger studies, see Günther (1908–1909), Réthly (1960), Varga (2008) and Kozák and Prachař (2010); the Mór map is reproduced in Fig. 2.1.

After 1800, the second earthquake map approaching later macroseismic maps was created by German mathematician P.N.C. Egen, who analyzed the Dutch earthquake of Feb 23, 1828 (Egen 1828).

Also the map by P. Egen is far behind the present macroseismic maps' standard. The author used an ordinary geographical map as a cartographic base, in which he underlined individual seismically afflicted settlements by different colors, corresponding to three couples of degrees of macroseismic intensity (he alone formulated his 6-degree macroseismic intensity scale): namely, the degrees 6 and 5 he underlined red, degrees 4 and 3 blue, and degrees 2 and 1 yellow. Isoseismal lines were not plotted in the map. However, in transalpine Europe P. Egen's map is the first in which an attempt is made to depict the earthquake intensity variations throughout the disturbed area (Davison 1978).

As the third example we mention the macroseismic intensity map presented by German naturalist, professor of mineralogy and mining in the University of Bonn, J.J. Nöggerath. His map showed the distribution of the macroseismic effects of the (Mid) Rhenish earthquake of 1846, Jul 29 (Noggerath 1847).

A quick look onto the map will confirm that the author utilized local reports on the earthquake for the construction of three isolines. The area of the largest damage reported was closed inside a small circle containing the towns Coblentz and St. Goar, inside of which also the highest intensity of the shock was felt. The next line of polygonal shape joins the outermost places at which the shock was actually felt. The third line, consisting of a dotted circle line, which surrounds the disturbed localities indicates the probable size of the disturbed area "So far I know the earthquake map contains the first attempt to draw isoseismal lines… leading to the first determination of the position of the epicenter by means of such lines" [cited from the text by Davison (1978)].

The presented examples bear witness how toilsome was the way towards the standard macro seismic intensity maps in the course of the first decades of the 19th century. In this gray area of slow progress all the more shines the remarkable advancement of O. Volger in the macroseismic mapping of the 1855 Visp event (Kozák and Vaněk 2006); his ideas gave a lead towards a meaningful and reasonable macroseismic earthquake analysis, as demonstrated in Sect. 2.2 of this text. Let us look in which degree these new achievements, including seismic cartography, were utilized in the analyses of South Poland earthquakes.

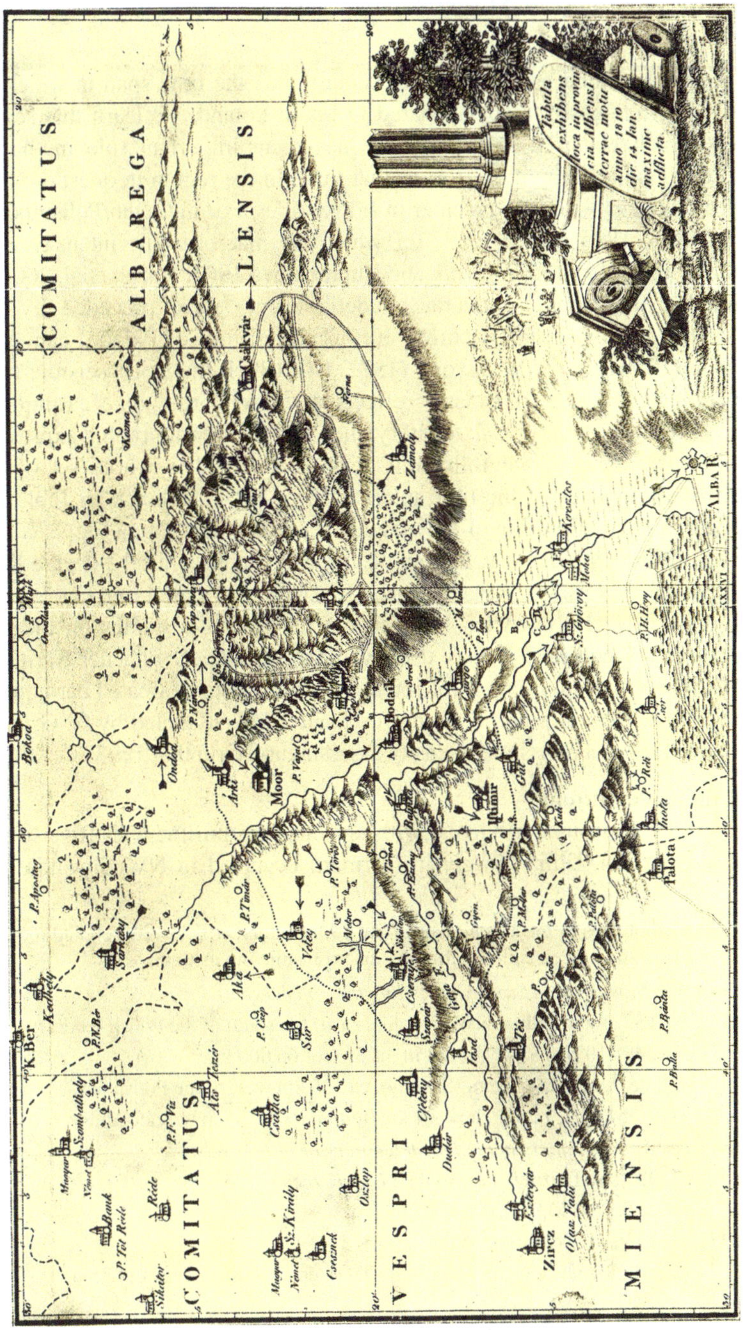

Fig. 2.1 The first MSM—by Tomtsanyi and Kitaibel—is related to the 1810, Jan 14, Mór (Hungary) earthquake

2.4 Selected Earthquakes in South Poland and Its Vicinity and the Respective Macroseismic Maps

By examining the second half of the 19th century, i.e., the time span in which the first macroseismic intensity maps appeared in South Poland, we learn that several earthquakes which occurred in this region played an important role in the all-European "earthquake mapping" effort of that time. In the region in question, three moderate earthquakes with the epicenter in Lower Silesia or in Czech/Polish border occurred in 1883, 1895 and in 1901, for which the macroseismic intensity maps were constructed (Réthly 1952; Kárník and Ruprechtová 1963; Kozák et al. 2002).

Additionally, these three moderate Sudetic events were preceded by two stronger seismic events, which occurred out of the territory of Poland, namely in NW Slovakia (1858) and in N. Saxony (1872), for which the macroseismic maps were also prepared. Since the seismic effects of these border-zone earthquakes were clearly felt deep in the Polish territory, they and their cartographic expression will also be mentioned in the following considerations. The whole set of five above-named earthquakes of the time interval 1858–1901 is shown in Table 2.1; for the geographical distribution of the treated events, see the map in Fig. 2.2.

At the end of the 19th century, in many European countries the 12-degree Rossi or Rossi-Forel macro-seismic intensity scales (RS, RFS) were used; see Forel (1881) and Rossi and Forel (1881). After 1964, in the retrospective studies on the 19th century Sudetic earthquakes, it was usually the Medvedev-Sponheuer-Kárník-64 scale (=MKS-64 scale) that served for intensity assessments of treated earthquakes (Medvedev et al. 1965); at present, the newest scale EMS-98 edited by G. Grünthal is recommended for macroseismic intensity assessments (Grünthal 1998).

The 1858 Žilina earthquake

On Jan 15, 1858, a strong earthquake with epicenter within the Váh River elbow near the present town Žilina (*Sillein* in German) occurred in NW Slovakia. The

Table 2.1 The earthquakes of the second half of the 19th century, which were felt and recorded in the Polish territory and processed in a form of a macroseismic intensity maps

Two earthquakes in Slovakia and Saxony preceding the Polish events
1858, Jan 15, 20^H 15^M, near Žilina, NW Slovakia, ca 40 km south of the Polish border
$I_0 = 9$ (MCS), epic. 49° 10′N, 18° 47′°E, felt in the area of 66,000 km^2
1872, Mar 06, 15^H 55^M, at Posterstein, (Saxony), ca 135 km west of the Polish border
$I_0 = 7.5$ (MCS), epic. 51.6°N, 12.3°E, h = 9.4 km, felt in: r = 150 km (circle radius)
Three earthquakes with epicenters in the Polish territory or in Polish border
1883, Jan 31, 14^H 43^M, at Trutnov, near to the Polish/Czech border
$I_0 = 7$ (MSK), epic. 50.5°N, 15.9° E, felt in 4,500 km^2
1895, Jun 11, 09^H 27^M, at Strzelin in Mid-Silesia
$I_0 = 7$ (MSK) and (MCS) and $I_0 = (6°–7°$ in R-F Scale), epic. 50.8°N, 17.0°E, felt in 25,000 km^2
1901, Jan 10, 02^H 30^M (GMT), at the Upper Úpa Valley, near to the Czech/Polish border
$I_0 = 7$ (MSK), epic. 50.5°N, 16.1°E, felt in 50,000 km^2

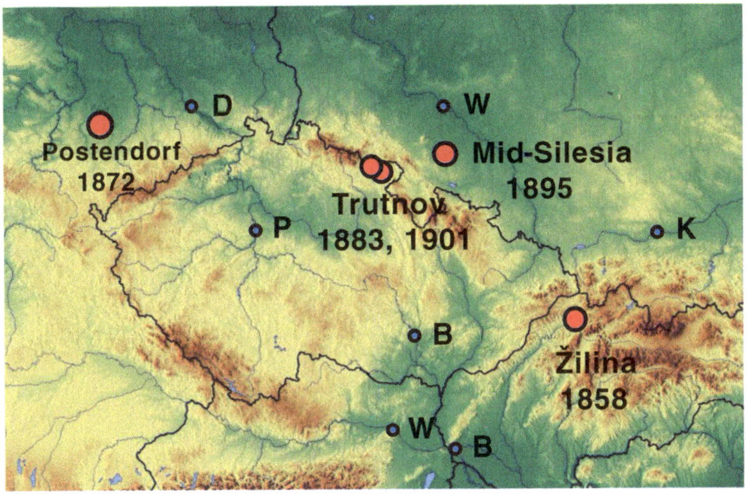

Fig. 2.2 Modern map showing the locations of the 19th century transalpineearthquakes discussed in this paper. B–Brno (CZ), B–Bratislava (SK), D–Dresden, K–Kraków, P–Praha, W–Wien (A), W–Wrocław (PL)

radiation pattern of the source displayed three maxima, towards S (up to Esztergom in Hungary), towards NW (up to Wrocław in Lower Silesia) and W (towards Moravia and Bohemia). The largest seismic damage was ascribed to the Žilina town where as many as 29 houses were damaged or ruined (classified as not safe for use). In the next 10 months of 1855, 16 aftershocks were recorded in the region.

This strong earthquake, the effects of which were clearly felt in the vast, densely inhabited adjacent regions, attracted large attention and interest of local naturalists and geo-savants, which resulted in numerous written reports; three of them were complemented by macroseismic maps. These maps carry some elements of novelty and therefore—as a whole—they also represent a kind of milestones in Central European macroseismic cartography. The two macroseismic maps in question, constructed by J.F.J. Schmidt, were attached to his 72-page report on the event (Schmidt 1858). In his first detailed map, reproduced in Fig. 2.3, the shape, orientation and size of the largest damage zone was outlined as an oval elongated in NW-SE direction, similarly as another outer oval which limits the region of weaker earthquake effects. The map is plotted in the network of geographical coordinates and, besides the river pattern, also main mountain ridges are given. The direction of seismic wave impact was recorded for some localities, see the 11 arrows in the inner zone, 13 ones in the outer zone. In the second, general map (see Fig. 2.4), a larger geographical segment of the affected region is given: the complementary macroseismic data enabled the author to construct the shape and orientation of three intensity zones between the Danube River and Lower Silesia and to distinguish them by different shades of red color. The map also gives three calculated circular isochrones: I, II, III.

Fig. 2.3 The MSM—by J.F.J. Schmidt—is related to the 1858, Jan 15, Žilina earthquake, a detailed map

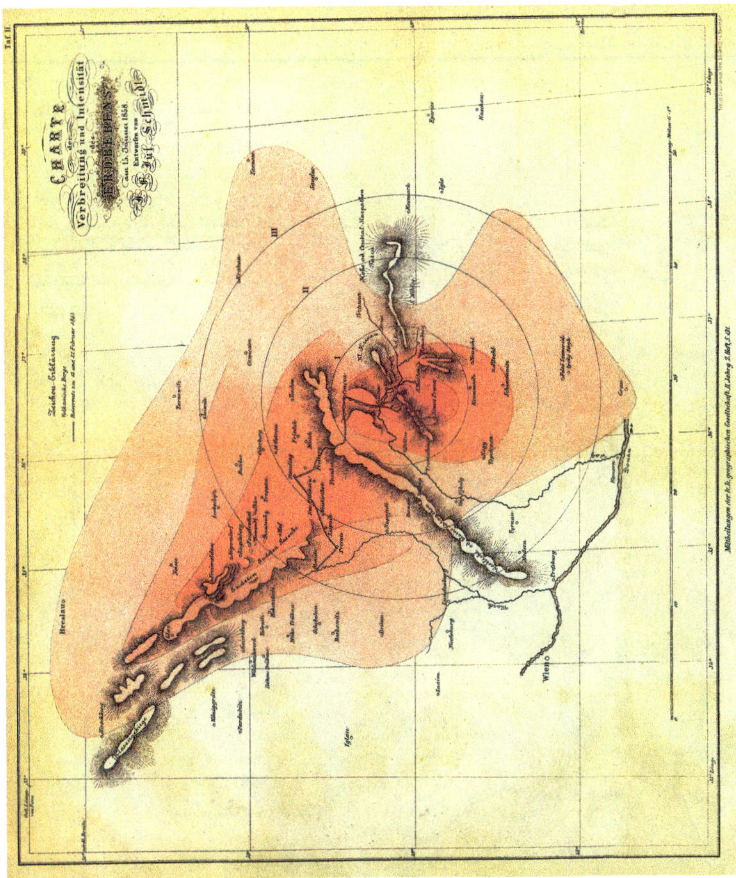

Fig. 2.4 The MSM—by J.F.J. Schmidt—is related to the 1858, Jan 15, Zilina earthquake, general map. © Jan Kozák, NKC Prague 2013, all rights reserved

The second naturalist who plotted a macroseismic map of the effects caused by the 1858 Žilina earthquake was L.H. Jeitteles (born in Prague), professor of biology, botany and geography; in 1858 he was professor at the grammar school in the Moravian Opava town. Shortly after the earthquake he visited Žilina and its broad vicinity, collected personally and in detail macroseismic effects and published them in a brilliant paper complemented by a color-lithographic macroseismic map reproduced in Fig. 2.5 (Jeitteles 1859, 1860), showing three zones corresponding to three levels of seismic damage, namely *small*, *middle*, and *maximum*. However, in contrast to the J.F.J. Schmidt's map in Fig. 2.4, the map by L.H. Jeitteles (also drawn

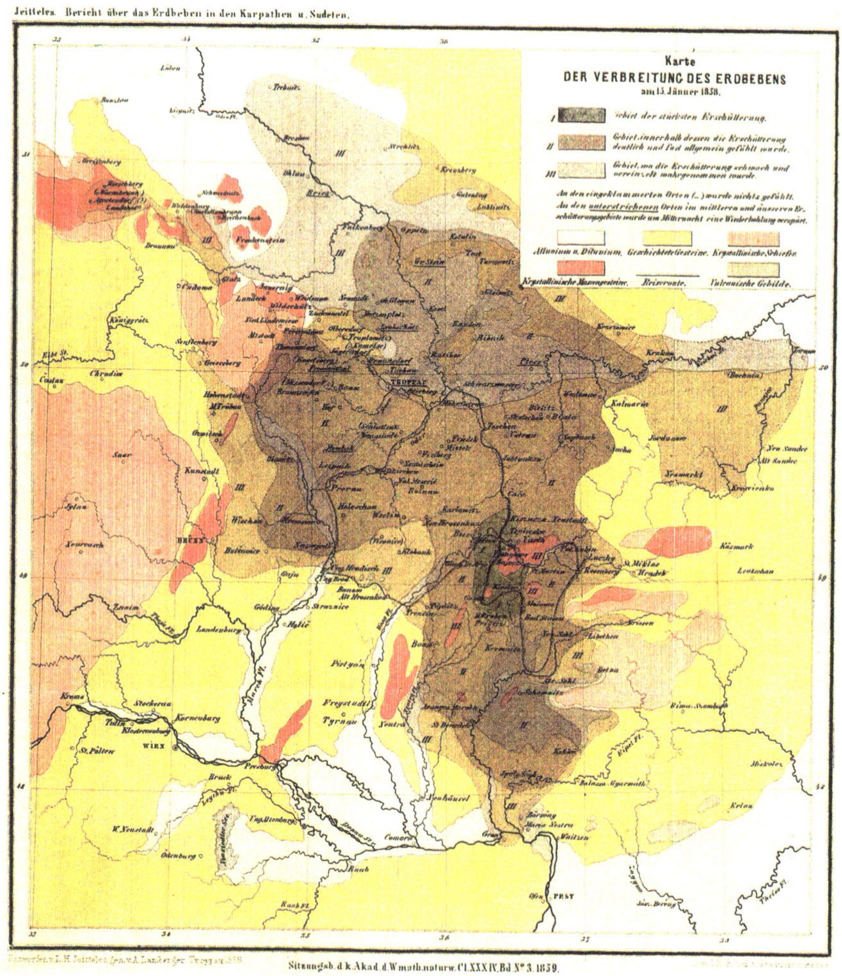

Fig. 2.5 The MSM—by L.H. Jeitteles—is related to the 1858, Jan 15, Žilina earthquake; seismic data are laid on geological backround.

with geographic coordinates and also prepared for the large region, from 47°N up to 51½°N) gives a much more detailed pattern of the three zones of macroseismic intensity. He was the first map maker who superimposed macroseismic data over the geological background. Thus, according to our knowledge, his map was the first in the world where seismic and geological fields were plotted together and their mutual relation discussed. In this sense the map by L.H. Jeitteles—no doubt—belongs to the milestones of world macroseismic cartography. The map was reproduced and discussed in detail by Vaněk and Kozák (2007).

Two further maps of the 1858 Žilina earthquake were prepared by M. Sadebeck, professor in Technical School in Opava. In his first macroseismic map-sketch, see Fig. 2.6, he gave the names of localities where the earthquake was felt, and also localized the epicenter zone. He demonstrated that the seismic effects were felt far in the Polish territory, from Nowy Sącz, Tarnów and Kraków in the northeast up to Wrocław, Kłodzko and Jelenia Góra in the northwest. Sadebeck, similarly as Schmidt and Jeitteles, confirmed that the general shape of the seismically loaded field was prolonged in the NW-SE direction (Sadebeck 1858).

In his second map (Fig. 2.7) he plotted a detailed map of the Žilina Valley (with 34 localities where seismic effects were reported) together with a remarkable city map of the Žilina town center, where the 29 damaged houses were marked, which was a novelty in macroseismic cartography.

The 1872 Mid-German Earthquake

The next strong earthquake occurred in the eastern part of Germany, in Saxony, on Mar 6, 1872, near to Posterstein by Gera, usually named *Leipzig earthquake* or *Middle German earthquake*. The principal investigation of this seismic event has been carried out by K.A.L. von Seebach, professor of geology at the university of Göttingen, inventor and promoter of the terms *isoseist*, *homoseist* and *pleistoseist zone* (Seebach 1873). Appreciating the map constructed by K. Seebach, reproduced in Fig. 2.8, we may state that he plotted the actually observed macroseismic intensity values as isolines which have the E ↔ W prolonged, far not circular shape (Seebach 1873; Sponheuer 1952).

Another map showing the shaken area of the earthquake was composed and presented in 1872 by L. Grebe, *Oberlehrer* (=deputy head teacher) in a grammar school in Cassel. In his presentation he summarized some of the earthquake effects and illustrated them in his map of German territories and adjacent parts of Bohemia, Bavaria and Poland, namely in Budziszyn, Lwówek Śl., Lubin, Wałbrzych, Głogów and Wrocław. However, Grebe in his map did not distinguish seismically affected localities according to degree of seismic damage so that he could not draw any isoseismal lines; that is why his map cannot be considered as a macroseismic map (for Grebe 1872 see Grünthal 1992).

The 1883 Trutnov earthquake

Cartographic portrayal of the extension of the 1883, Jan 31, Trutnov (*Trautenau* in German) earthquake in the northeast Bohemia was presented by two naturalists, G.C. Laube of the Austro-Hungarian monarchy and H. Kunisch of Prussian Silesia's Wrocław (*Breslau* in German). Laube (1883) in his map reproduced in Fig. 2.9

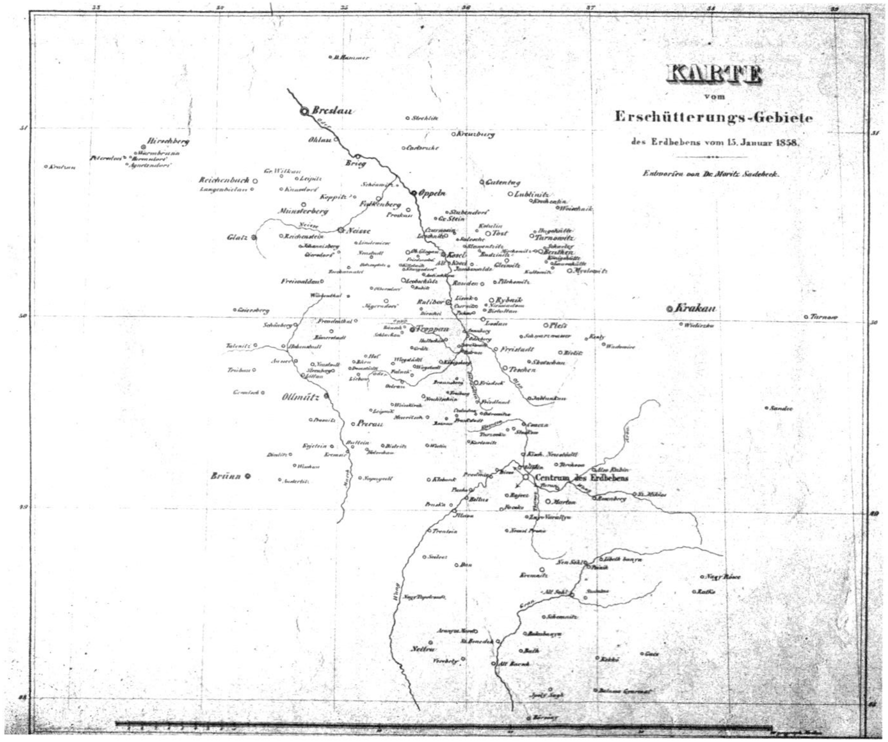

Fig. 2.6 The MSM—by M. Sadebeck—is related to the 1858, Jan 15, Žilina earthquake. Affected zone and localities are plotted.

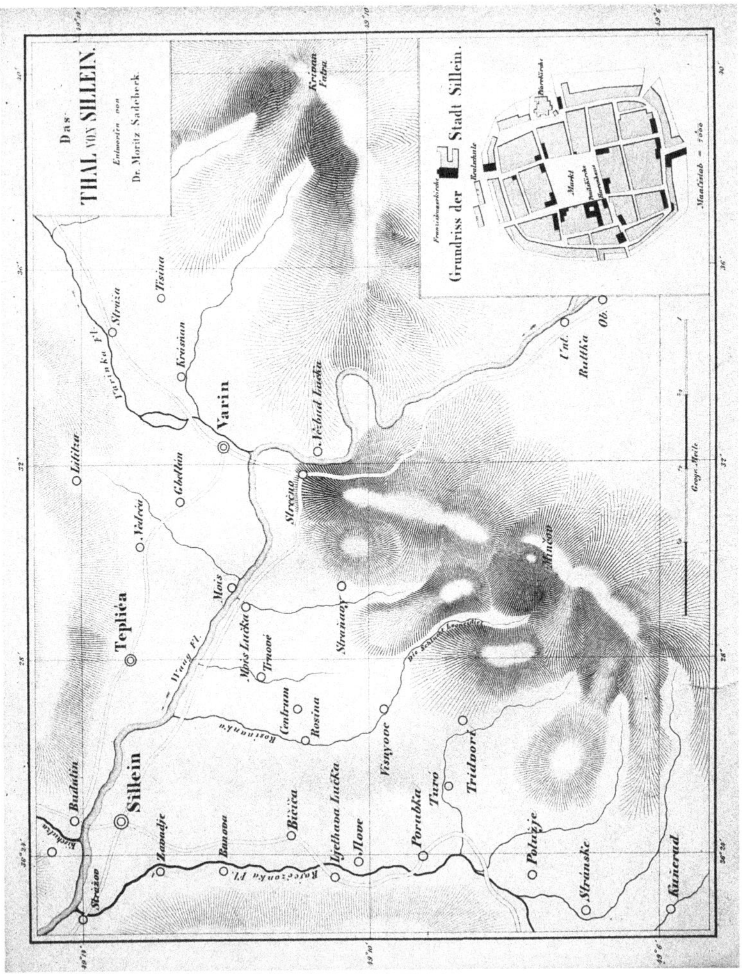

Fig. 2.7 The MSM „Das Thal von Silein" was made by M. Sadebeck. The map is related to the earthquake of Jan 15, 1858. Small marginal city map shows the edifices damaged by the earthquake.

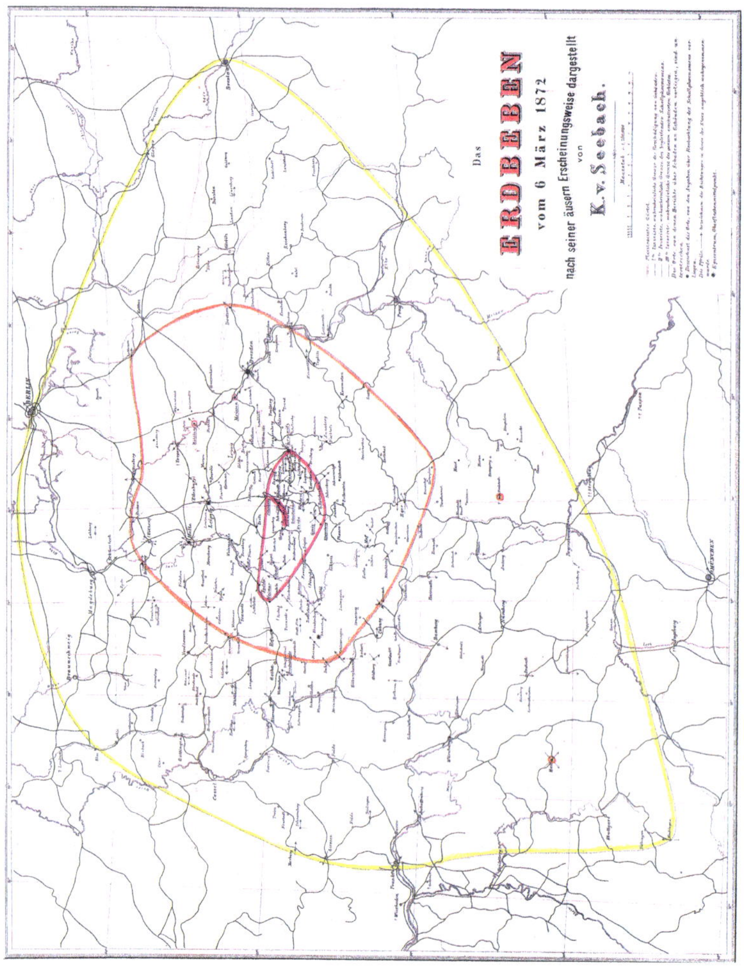

Fig. 2.8 The MSM—by K. von Seebach—is related to the 1872, Mar 12, *Mid-Deutsche-Erdbeben*. Note four *isoseismal lines*

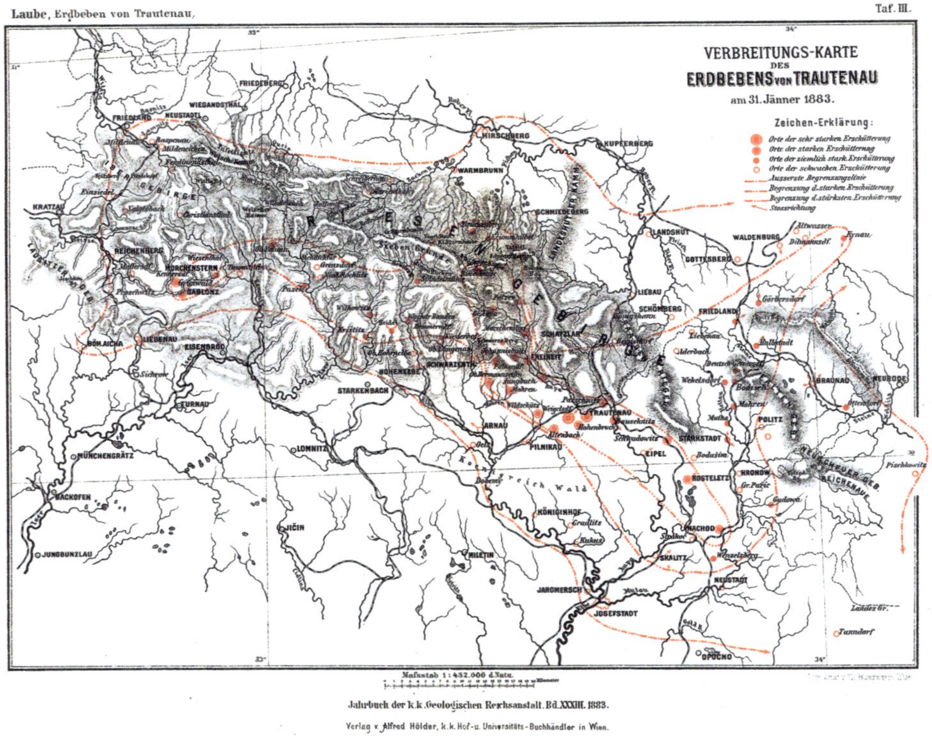

Fig. 2.9 The MSM—by G. Laube—is related to the 1883, Jan 31, Trutnov earthquake. Note the unusual shape of pleistoseismal zone; 3 *isoseismal lines* are plotted in the map

classified the affected localities in four levels of seismic intensity (very strong, strong, rather strong and low); for some localities he marked directions of seismic impact. He utilized local reports for constructing three isoseismal lines. As for the first zone (which may represent the pleistoseist zone), its extraordinarily curious shape should be pointed out: it has a S-shaped form of a narrow strip of a length of 50–52 km and width of 5–6 km, see the map in Fig. 2.9. It should be noticed that the zone of the highest macroseismic intensity, between the Sněžka Mt. towards the Náchod town, lies entirely on the territory of Bohemia. The whole prolonged macroseismic field is predominantly oriented as parallel with the Krkonoše main mountain ridge, i.e., in the NW-SE direction.

In the map shown in Fig. 2.10, which was created by Silesian seismologist H. Kunisch, on the other hand, only two zones are plotted. The inner zone, non-regularly lobed, extending from SE to NW, about 100 km in length and 20–30 km in width, is erroneously denoted as *pleistoseismal* zone. The second dashed line gives "approximate limits of the earthquake", the limit line of the earthquake effects' perception. Localities with negative (or no) reports are also marked. In contrast to the orientation of the pleistoseismal zone laid in the NW-SE direction, the line denoting the limit of ground shaking has its axis of prolongation considerably inclined to the north, *due to a single point in the map sketch, namely due to positive reports from the Żagań town.* Evidently, without this single „Żagań" point (surrounded by localities reporting negative observations) the perception zone would be radically reduced in size and re-oriented similarly as the pleistoseismal zone (Kunisch 1883). We may conclude that the map by H. Kunisch was struck by several errors.

The 1895 Mid-Silesia earthquake

At the turn of the spring of 1895, a strong earthquake occurred in Lower Silesia on Jun 11th. Three authors portrayed the event cartographically. First, R. Leonhard and W. Volz, presented two specially drawn maps (see Figs. 2.11 and 2.12) of this earthquake and published them in Wrocław in 1895 (first version) and in Berlin in 1896 (second, complemented version). In these maps, the isoseismal lines separating the zones of macroseismic intensities corresponding to the 3rd, 4th, …, 7th degrees of the Rossi-Forel intensity scale were plotted, together with the isochrones and seismic impact directions (*Stossrichtungen* in German). Both their *Übersichts-Karten* must be appreciated as good-standard macroseismic maps of the 1890s. In Fig. 2.12 the second, supplemented map version is shown; the isoseismal lines are plotted over geological background, similarly as in the map by L.H. Jeitteles (see Fig. 2.5), related to the 1858 Žilina earthquake. In small left bottom corner-map a comparison is given of the extension of the 1895 earthquake (wrongly dated as of June 15th instead of June 11th), with two significant seismic events, which were felt in Lower Silesia, namely the 1883 Trutnov and 1858 Žilina earthquakes. The detailed and very carefully prepared map enables to identify two separate zones of maximum intensities, namely Strzelin-Münsterberg and Pieszyce-Diersdorf, between which the seismogenic fault can be traced by means of the seismic impact directions given between both the regions by arrows (Leonhard and Volz 1895, 1896).

The third map, shown in Fig. 2.13, also related to the 1895 Middle Silesian earthquake, was prepared by E. Dathe; for his map he used large, standard

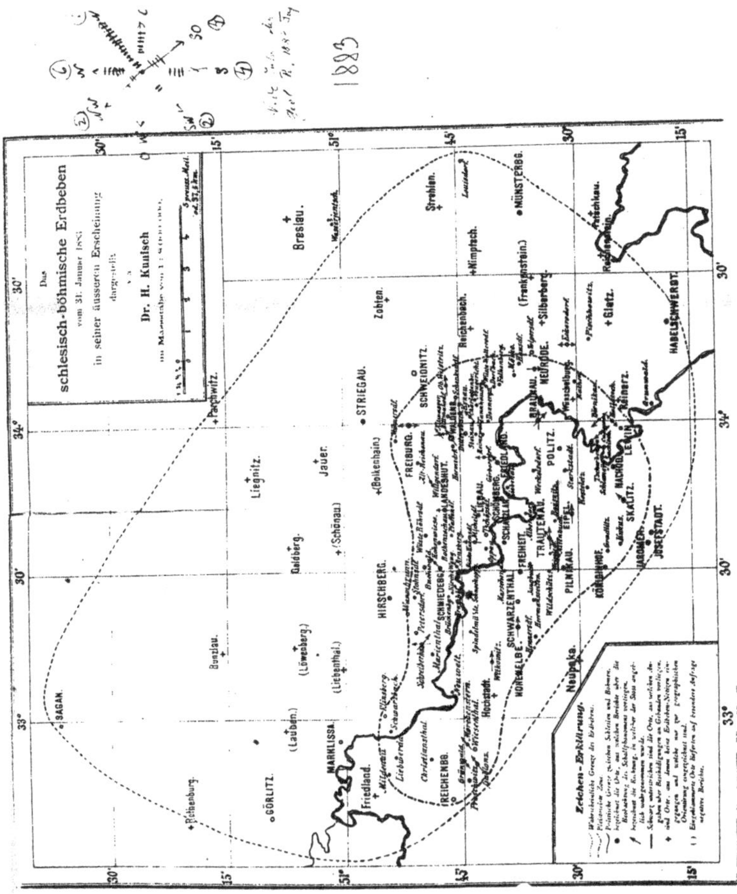

Fig. 2.10 The MSM—by H. Kunish—is related to the 1883, Jan 31, Trutnov earthquake. Note the unusual size and shape of the last *isoseismal line* going far to north due to single locality "Sagan" lying inside the field of negative reports

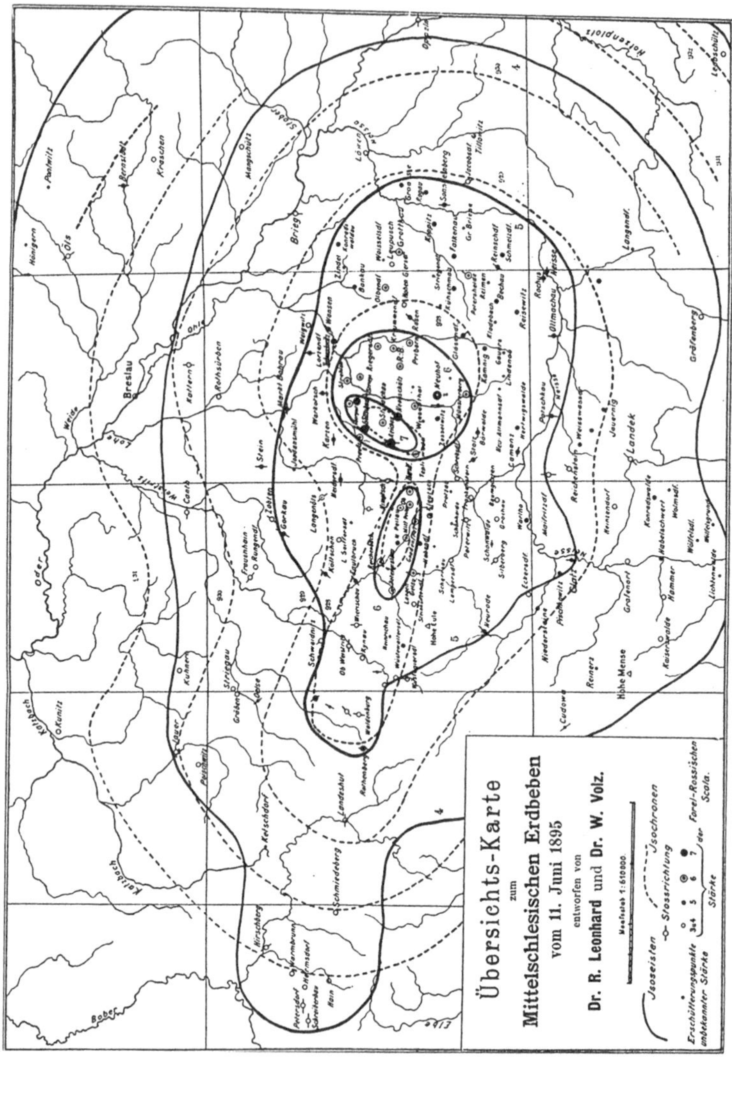

Fig. 2.11 Preliminary MSM—by R.V. Leonard and W. Volz—is related to the Mid-Silesian 1895, Jun 11, earthquake. Both *isoseismal lines* and *isochrones* are plotted. © Jan Kozák, NKC Prague 2013, all rights reserved

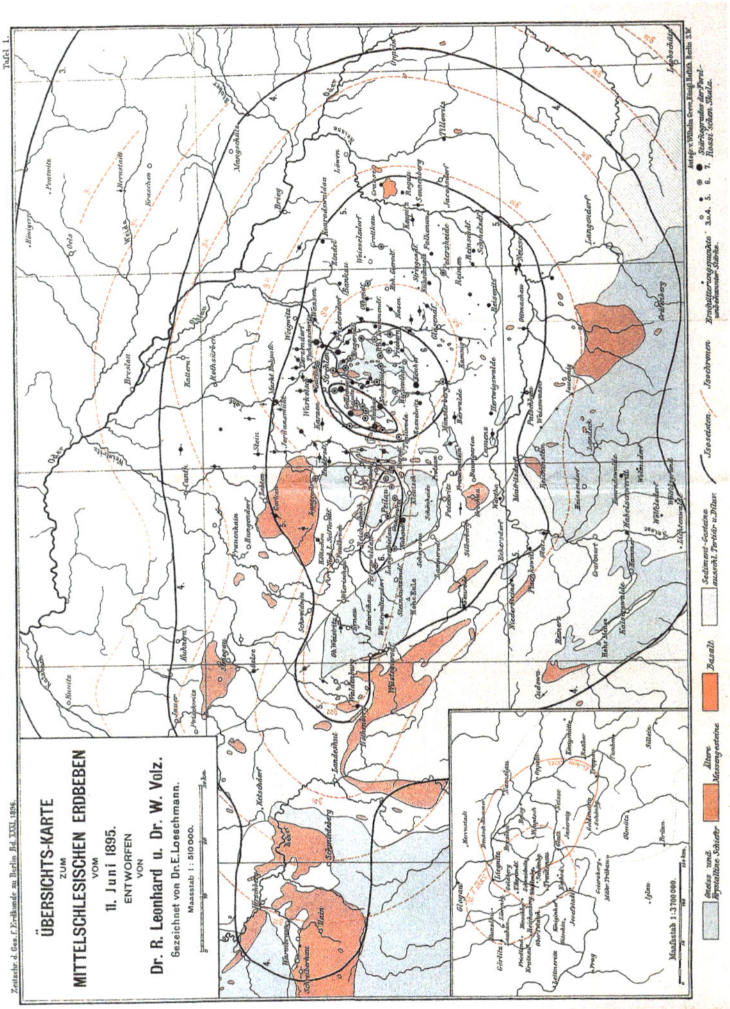

Fig. 2.12 Complete MSM—by R.V. Leonard and W. Volz—is related to the 1895, Jun 15, Mid-Silesian earthquake. Simple geology, intensity classification after Rossi and marginal map of two other earthquakes are given.

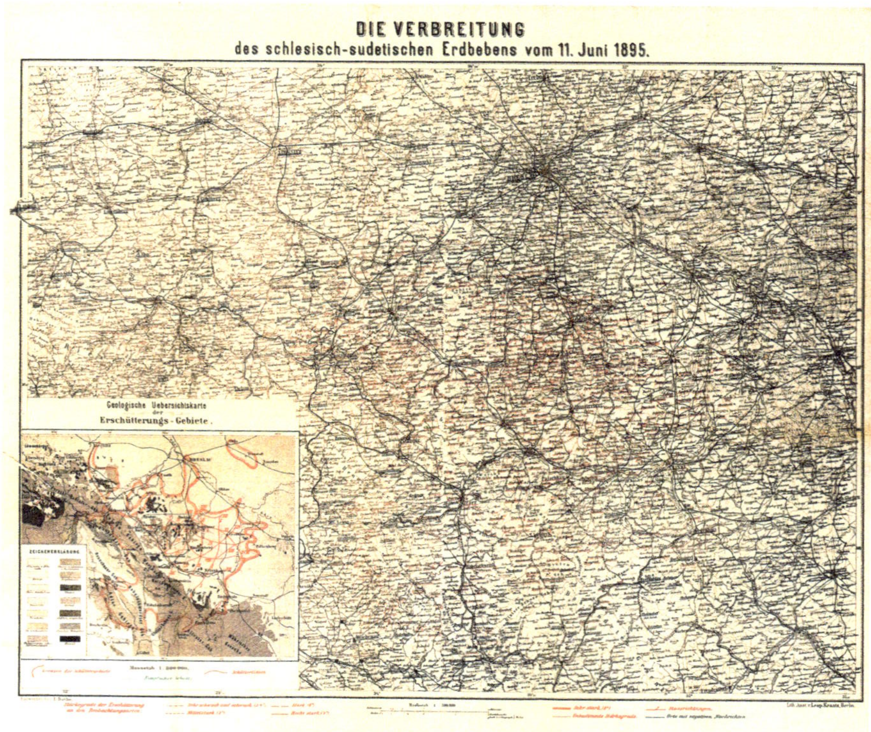

Fig. 2.13 The MSM—by F. Dathe—is related to the 1895, Jun 15, Mid–Silesian–Erdbeben. The marginal map given in the *left bottom corner* shows local geology and limits of perception zone

geographical map of Lower Silesia, into which he plotted various red-color marks, underlined the locality names and inserted other symbols for classification of the affected localities in 5 levels of earthquake intensity degrees, 3°–7°, of not specified intensity scale (namely very weak and weak, mid strong, strong, very strong, extremely strong). The large main map—already filled densely with standard cartographic information—after addition of numerous macroseismic symbols—became practically unreadable, especially when the map is reduced in size. The main map, however, is complemented with a small additional map of the region, where, on the background of the local geology (12 geological structures), geographical limits of earlier earthquakes and positions of principal faults are given. In general, E. Dathe's map carries a great amount of valuable macroseismic data; however, a magnifying glass is necessary (Dathe 1897).

In the last years of the 19th century, most papers analyzing individual weak and middle size earthquakes which occurred in the Central European region, were complemented with larger or smaller macroseismic maps. Not all of them, however, reached the high standard and quality of the maps by L.H. Jeitteles (Fig. 2.5), G. Laube (Fig. 2.9) and C. Leonhard and W. Volz (Fig. 2.12) discussed above. On the contrary, many of them appeared as simple map sketches, especially those preparet for weak seismic events.

The 1901 Sudetic earthquake

The Czech naturalist N.J. Woldřich, one of the authors of the 1901 event evaluation, named systematically this earthquake as "*The 1901 North-East Bohemian Earthquake*", since most of its pleistoseismal area occurred in Bohemia (=Czech Republic at present), similarly as the previous 1883 Trutnov event. However, the German speaking naturalists named this earthquake as „*The 1901 Sudetic earthquake*". Having in mind that the „Sudetic" area lies in Polish as well as in Czech foothills of Krkonoše Mts., the German name seems to be correct as well.

J.N. Woldřich published two papers on the 1901 event, first in Czech and repeatedly also in German academic journals; both were complemented by two macroseismic maps; they are reproduced here as Figs. 2.14 and 2.15. In the map in Fig. 2.14 he denoted all the towns, villages and other settlements afflicted by the earthquakes, according to the Rossi scale (degrees: 7°, 5°–6°, 4° and 3°); he even distinguished the localities in which the displacement was accompanied by underground sounds. In the map, also principal parallel fault zones were drawn. The isoseismal lines were not plotted, except of the shape and size of the epicenter zone. In the second map (Fig. 2.15), also originally published in the Czech paper (Woldřich 1901a, b), the general situation illustrating the afflicted area was given: seismic intensities are classified in three levels according to the Rossi scale and mutually contrasted by the three tunes of blue color. It is apparent from the map that the seismic field in Lower Silesia, Upper and Lower Lusatia and in eastern Saxony is clearly prolonged towards NW, from Frývaldov in the Moravian part of Silesia up to Leipzig in Saxony. The paper published in the Austrian academic journal (Woldřich 1901b) carries the same pair of macroseismic maps as the Czech paper and therefore it will not be discussed here.

Fig. 2.14 The MSM—by J.N. Woldřich—is related to the 1901, Jan 10, northeast Bohemia earthquake. Pleistoseismal zone and 3 main faults (in SE–NW direction) are given by the *red lines*. © Jan Kozák, NKC Prague 2013, all rights reserved

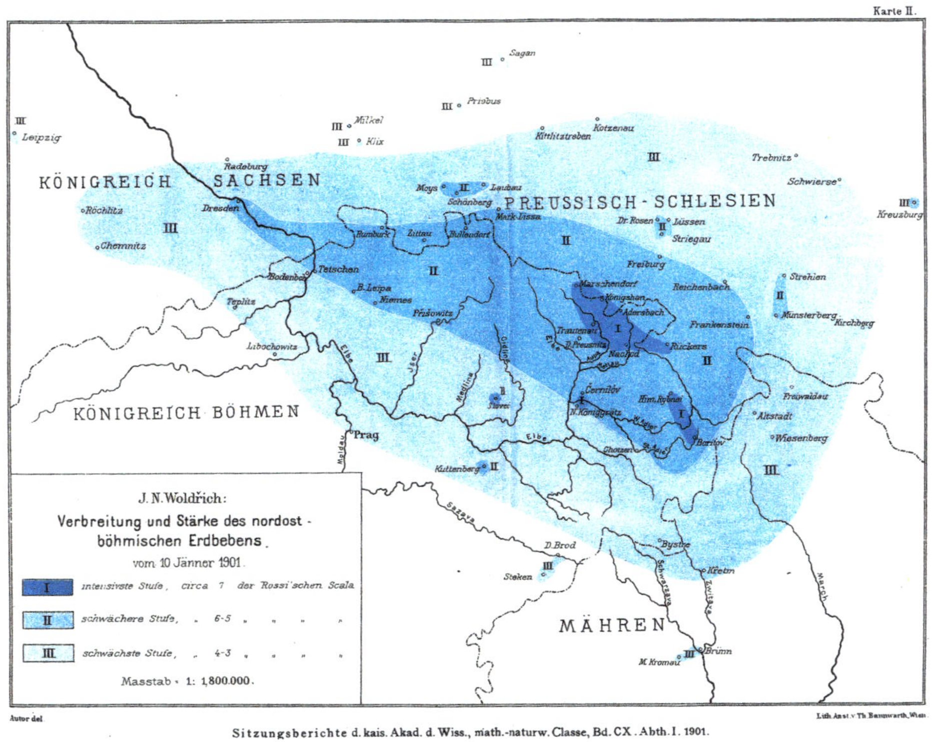

Fig. 2.15 The MSM—by J.N. Woldřich—is related to the 1901, Jan 10, in northeast Bohemia. Three intensity levels are differentiated by three tunes of *blue*.

The map by H. Credner, see Fig. 2.16, represents one-color (blue) macroseismic map, which neither cartographically nor seismically surpasses the common standard of the time. Two tunes correspond to the degrees of Rossi intensity scale, (5°–6°)—darker blue and (4°–5°)—lighter blue. By a red line, the Great Lusatia Dislocation runs in the central part of the seismically afflicted field from SE to NW. In the map margin there are also distant localities shown, from which positive reports also arrived. The map is not easily readable due to intensive blue tune shading, especially the zone of strong seismic effects (Credner 1901).

The map by J. Gränzer, see Fig. 2.17, who has seen the earthquake effects from the Czech part of the *Sudetes*, resembles the map by J.N. Woldřich given in Fig. 2.14. The author classified the intensity degrees in four levels on macroseismic effects while the fifth level was reserved for the localities declaring negative reports. The Great Lusatia Dislocation is given in the map, as a straight line broken in the Velká Úpa town lying on the Úpa River in NE Bohemia. By four variable tunes of grey color, four geological structures are denoted: Primary and Silurian, Bitumen, Permian, Cretaceous. The map carries author's seismic sketch of a cross-section of the Earth (Gränzer 1901).

The last couple of maps related to the 1901 Sudetic earthquake was prepared and published by F. Sturm, who was active in the Geological Institute of the Wrocław University. His first (general) map, see Fig. 2.18, shows the seismically loaded field plotted for the 1901 earthquake prolonged between Magdeburg and Moravská Třebová. Except of several details, the Sturm map is practically identical with the map by J.N. Woldřich, see Fig. 2.14 The seismic intensity is classified in three degrees of the Rossi intensity scale (7°, 6°–4° and 3°–4°). Quite evident is the NW-SE orientation of the seismic field, which is stretched up to distant Magdeburg in the map. The second Sturm's map, reproduced in Fig. 2.19, brings an interesting comparison of four earthquakes (Sturm 1903). In 1858 and 1901, Lower Silesia was vexed by two strong earthquakes and in 1883 and 1895 by two moderate seismic events. It is noteworthy that the three „Sudetic" earthquakes of 1883, 1895 and 1901 display nearly the same pleistoseismal zone of maximum intensity, between Jelenia Gora, Świdnica and Legnica. Can this area be understood as a zone of Lower Silesia earthquake foci concentration? In contrast to it the 1858 Žilina earthquake, the perceptivity limits of which reached the JelGóra-Świd-Leg zone, displayed a specific behavior: the seismic effects of this distant earthquake seemed to avoid the entrance into this „Sudetic earthquake foci center".

2.5 Poland: Maps of Seismicity?

Seismic events, which always meant a danger for population in seismically active regions, represented, on the other hand, a strong engine drafting local naturalists to search for the measures of the dreaded seismic effects' mitigation.

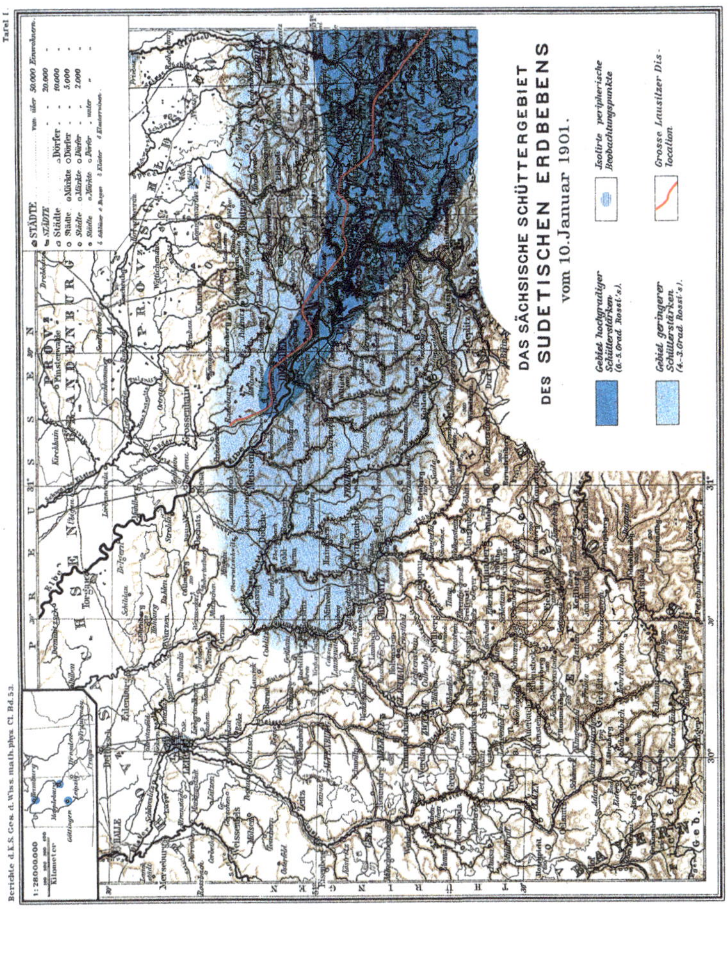

Fig. 2.16 The MSM—by H. Credner—is related to the 1901, Jan 10, northeast Bohemia earthquake. Three degrees of seismic intensity are expressed by three tunes of *blue* colour

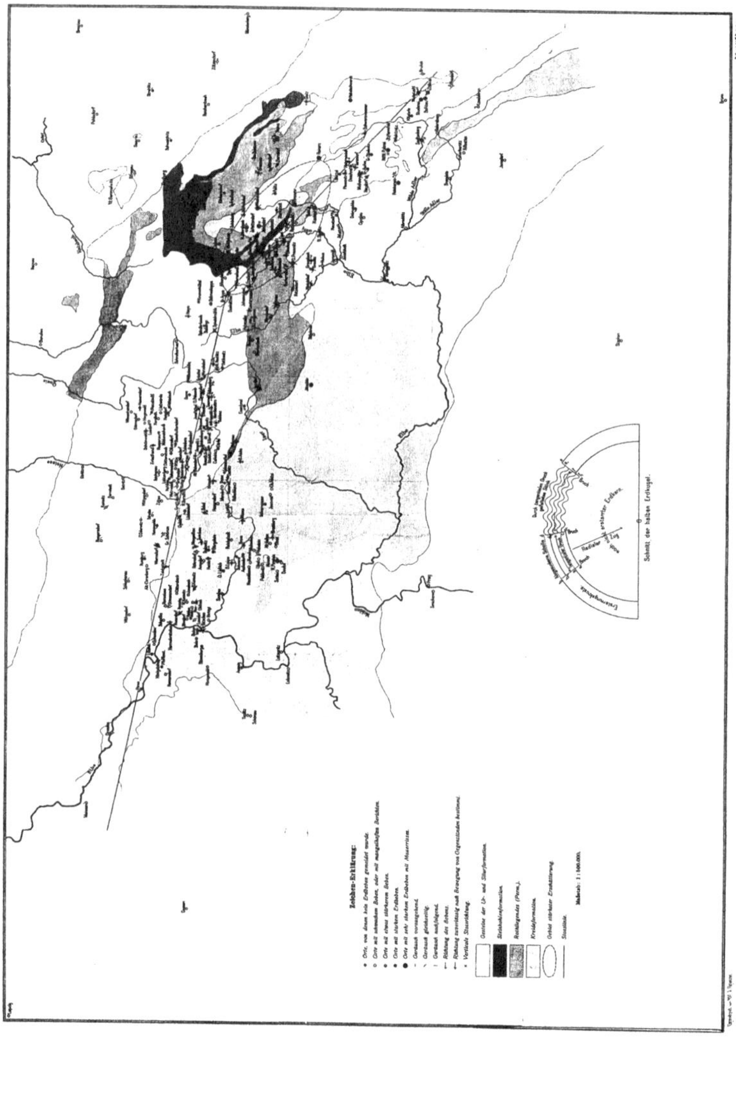

Fig. 2.17 The MSM—by J. Gränzer—is related to the 1901, Jan 10, northeast Bohemia earthquake. Seismic observations and geological structures together with geological faults, are plotted in the map

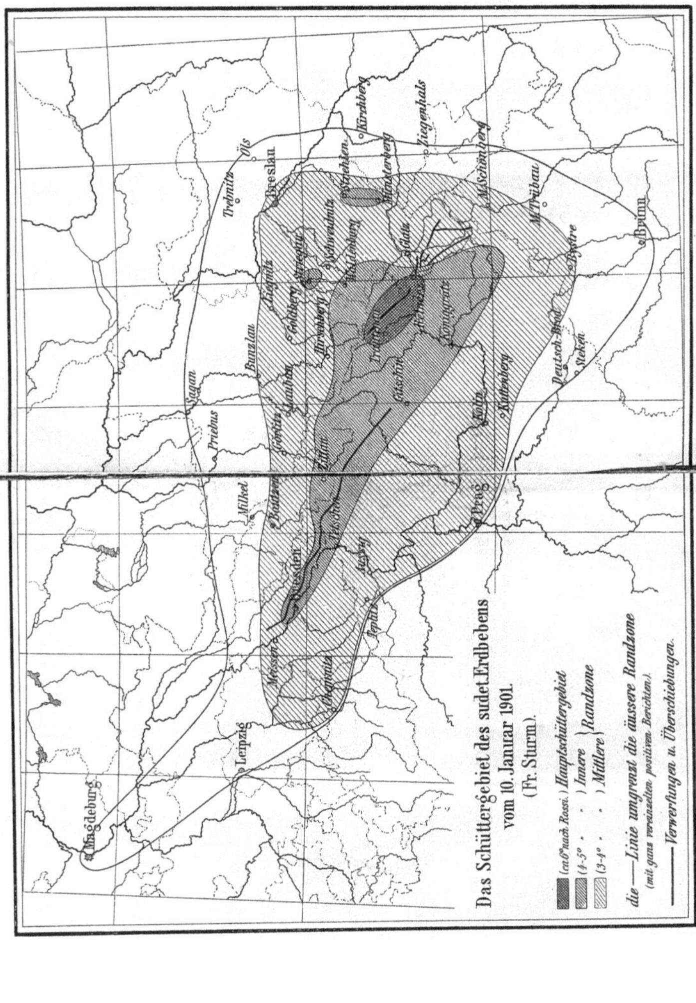

Fig. 2.18 The MSM—by F. Sturm—is related to the 1901, Jan 10, northeast Bohemia earthquake. *Isoseismal lines* separate 4 zones (levels of seismic intensity). The map resembles the map by Woldrich in Fig. 2.14. © Jan Kozák, NKC Prague 2013, all rights reserved

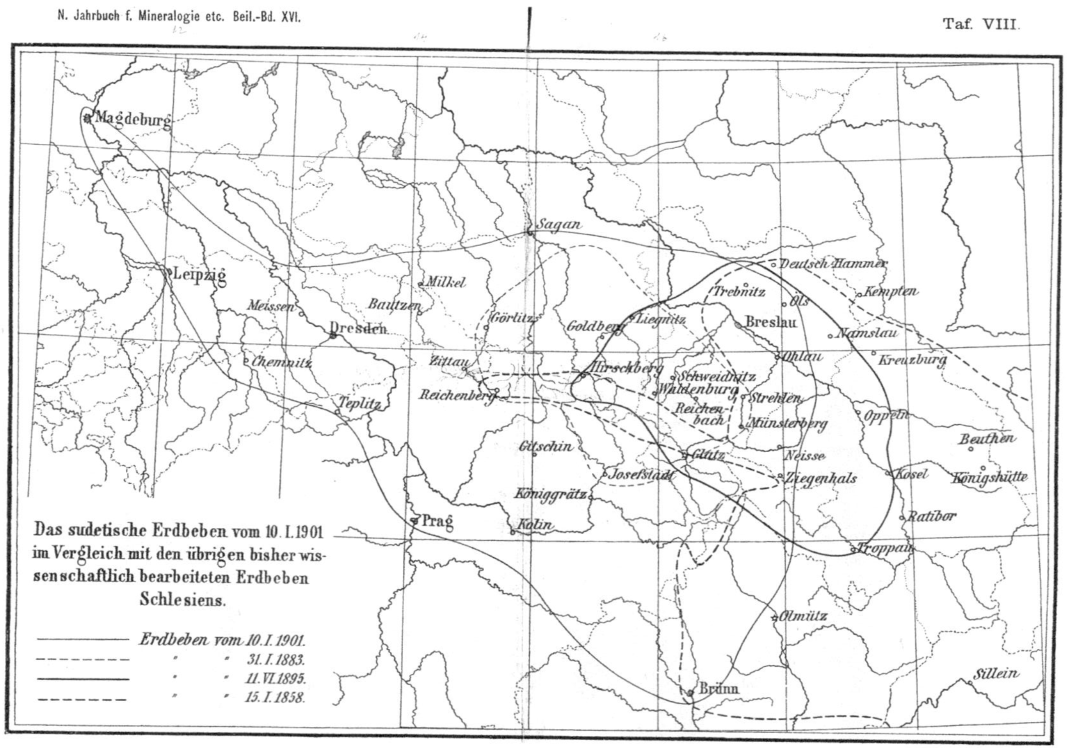

Fig. 2.19 Map of south Poland, north Bohemia and east Saxony: Limits of perception of 4 strong earthquakes, which were felt in the region in the years 1858, 1883, 1885 and 1901, are plotted and pertinent earthquakes mutually compared

And the macroseismic intensity maps seemed to be a kind of building blocks for such an effort.

However, even though nobody would query or oppose such a conclusion, the correct assessment of seismic danger has always been loaded by numerous circumstances and conditions, namely:

- by the length of the observational period, which should be long enough to enable collection of statistically sufficient number of seismic recordings. Evidently, the longer the observational period is the higher the degree of reliability may be attributed to the resulting pattern of seismic danger.
- by the occurrence of strong earthquakes, the epicenter and isoseismal lines of which would be more easy to determine.

Coming back to the assessment of the seismic danger in the area loaded by moderate seismic impact we have to agree that the above-discussed five seismic events felt and recorded in South Poland (for which macroseismic maps were prepared in 1858–1901) are not sufficient for the desired assessments. We may conclude that for these seismically relatively quiet regions (only 4 earthquakes occurred Poland having the intensities $I_o \geq 7°$ of Mercali-Cancani-Sieberg scale in the time span 1011–1970) the *macroseismic treatment* of the weaker seismic event does not seem to be an effective tool for the required achievement of the seismic danger assessment; see, e.g., the unequal classification of the “boundary” earthquakes—the 1783 Trutnov and the 1901 NE Bohemia ones, presented by the that time Czech, Polish, Silesian and German naturalists (Schweitzer, undated).

As concerns the seismically (moderately) active zone running approximately parallel along to the Czech-Polish border (sometimes called the Sudetic-West Carpathian zone), for making a reliable seismic danger assessment the seismologists would need the outcome of sensitive instrumental seismic monitoring performed in the country in the course of the 20th century, associated with seismic mapping of Poland.

2.6 Conclusions

As concerns the early macroseismic intensity maps of earthquakes constructed in the second half of the 19th century for Polish territory, and for Polish border regions we collected altogether 17 maps, related to 5 earthquakes.[3] It appeared that the largest obstacle in constructing these maps has laid in the lack of strong earthquakes occurrence; weak earthquakes and weak ground shaking during seismic swarms, which make the majority of seismic activities there, did not provide enough data necessary for constructing standard and reliable macroseismic maps.

[3] Six maps, namely those presented in Figs. 2.5, 2.6, 2.7, 2.14, 2.15 and 2.18, are deposited in private collection “New Kozak Collection” (*NKC*) and are protected by copyright Jan Kozák ©.

Regardless of this limitation, five medium-size earthquakes, VI $\leq I_0 \leq$ IX (MSK-64), which occurred in the regions of Žilina 1858, Gera 1862, Trutnov 1883, Mid-Silesia 1895, and the 1901 Sudetic event, were processed and the results were expressed in 17 excellent macroseismic maps. It is demonstrated in the text of the previous section that some of these 17 maps, mainly those prepared by J. Schmidt, L. Jeitteles, K. Seebach, G. Laube, C. Leonhard, W. Volz and J. Woldřich, may be considered as remarkable macroseismic maps of the period.

We may conclude that in the 19th century Central Europe became one of important world centers of earthquake macroseismic research, in which geo-savants active in Poland played an important role: In the last decades of the 19th century, numerous institutions for studying Earth sciences—including seismic manifestations—were established in Central Europe; Kraków, Ojców, Wrocław, Lwów and Królewiec/Kaliningrad definitely belong to them.

The proper methodology and technology of macroseismic mapping appeared as a necessary step towards constructing the long desired *maps of seismicity*, which today represent the essential manuals regulating construction of large, seismically endangered edifices in seismically active regions. This, however, is already the story of the 20th century.

Acknowledgments The author is indebted to Jiří Vaněk of the IG, Prague, for his stimulating suggestions and valuable comments on the paper text. To Petr Jedlička associated with the same Institute the author thanks for his technical assistance, especially as concerns processing of the reproduced maps.

References

Barbano MS, Cosentino M, Lombardo G, Patane G (1980) Isoseismal maps of Calabria and sicily earthquakes (Southern Italy), Catania. CNR-PFG (Consiglio Nazionale delle Ricerche, Progetto Finalizzato Geodinamica), Gruppo di Lavoror „Catalogo dei Terremoti", pub. No. 341, p 116+

Bertrand E (1757) Memories historiques et physiques sur les tremblement de terre. La Haye

Credner H (1901) Das Sachsische Schuttergebiet des Sudetischen Erdbebens vom 10. Januar 1901, Ber. d. math.-phys. Cl. d. k.k. Gesell. d. Wiss. zu Leipzig; Sitz. 4 Marz 1901, pp 83–103

Dathe E (1897) Das schlesisch-sudetische Erdbeben vom 11. Juni 1895, Abhandl. d.k. preuss. geol. Bundesanstalt, Neue Folge, Heft 22, Berlin

Davison C (1978) The founders of seismology. Arno Press, New York Times Company, New York

Dolomieu D (1785) Memoire sur les tremblements de terre de la Calabrie pendant l'annee 1783, A. Fulgoni, Roma

Egen PNC (1828) Über das Erdbeben in der Rhein- und Niederlanden vom 23. Februar 1828. Annalen der Physik und Chemie, Band 13(89):153+

Foglia GA (1627) Historico discorso del gran terremoto successo nel Regno Napoli…nel corrente anno 1627 a di 30 luglio…, Napoli

Forel FA (1881) Archives Scientifiques. Phys Nat Vol 6

Gränzer J (1901) Das sudetische Erdbeben vom 10. Jänner 1901, Mitteilungen des Vereines der Naturfreunde in Reichenberg, Jahrg. 32, Liberec (Reichenberg)

Grünthal G (1992) The central German earthquake of Mar 6, 1872. In: Gutdeutsch R, Grünthal G, Musson R (eds), Historical earthquakes in Central Europe, Abhandlungen d. geol, vol 48. Bundesanstalt, Wien, pp 51–109

Grünthal G (ed) (1998) European Macroseismic Scale 1998, Cahiers du Centre Europeen de Geodynamique et de Seismologie, vol 15, Luxembourg

Günther S (1909) Das westungarische Erdbeben des Jahres 1810. Natur und Kultur VI, Muenchen, pp 22–23

Jeitteles LH (1859) Bericht über das Erdbeben am 15. Jänner 1858 in Karpathen und Sudeten, Sitzungberichte der k. Akademie der Wissenschaften, mathematisch – naturwiss. Classe. Band 35, Nr. 7, Wien, (mit 1 Karte)

Jeitteles LH (1860) Versuch eine Geschichte der Erdbeben in der Karpathen – und Sudeten – Landern, pp 87–349

Kárník V, Ruprechtová L (1963) Seismicity of the Carpathian Ridge bis zu Ende des Achtzenten Jahrhunderts, Z. d. deutsch. Geol. Gesell. 12, 2, Acad Sci 182:143–187

Kárník V (1960) Die Seismizität der Westkarpathen. Travaux de l'Institute geophysique, No. 134, Prague, pp 265–283

Kárník V, Michal E, Molnár A (1957) Erdbeben Katalog der Tschechoslowakei bis zum Jahre 1956. Travaux de l'Institute Géophysique de l ´Academie des Sciences. No. 69, Geofyzikální sborník, Praha, pp 411–598

Kendrick TD (1956) The Lisbon earthquake. Methuen & Co. Ltd, London

Kitaibel P, Tomtsanyi A (1814) Dissertatio de terrae motu in genere, ac in specie Morensi, Anno 1810, die 14 Jannuarii orto. Budae, Typus Regiae Universitatis 1814, Akademiai Kiado, Budapest

Kozák J (2001) 100-year aniversary of the First international seismological conference. Studia Geod Geoph 45:200–209

Kozák J, Buben J, Vencovský M (2002) Delineation and segmentation of the Peripienian Lineament. Acta Geophysica Polonica 50(2):179–206

Kozák J, Prachař J (2010) First macroseismic maps in Transalpine Europe related to the 1810 Mór earthquake in the Panonian Basin. Studia Geoph Geod 54:339–341

Kozák J, Vaněk J (2006) The 1855 Visp (Switzerland) earthquake. Early attempts of earthquake intensity classification. Studia Geophys Geod 50:147–160

Kunisch H (1883) Das schlesisch-bohemische Erdeben vom 31. Januar 1883. Jahrber. d. schles. Gessel. f. vaterl. Kultur, geogr. Section XL, Breslau, pp 318–344

Láska V (1902) Die Erdbeben Polens. Mitt. d. Erdbeben-Comm. d. k. Akad. d. Wiss., N. f., No. 8, Wien

Laube GC (1883) Das Erdbeben von Trautenau am 31. Janner 1883. Jahrb. d. k. geol. Reichsanstalt in Wien, vol 33, Heft 2

Leonhard R, Volz W (1895) Das mittelschleisische Erdbeben vom 11. Juni 1895, Jahresbericht d. schles. Gesell. f. vaterl. Kultur, Naturw. section, II. Abt. XXIV., Breslau

Leonhard R, Volz W (1896) Das mittelschlesische Erbeben vom 11. Juni 1895, Zeitschrift f. Erdkunde zu Berlin, Bd. XXXI, pp 1–21

Leydecker G (2011) Erdbeben Katalog fuer Deutschland mit Randgebieten f. die Jahre 800 bis 2008, Geol. Jahrb., Reihe Geophysik, Heft 59, Hannover, pp 1–198

Mallet R (1863) Great Neapolitan earthquake of 1857, the first principles of observational semiology as developed in the report to the Royal Society of London, London

Medvedev SV, Sponheuer W, Karník V (1965) Seismic intensity scale—version 1964. Soviet Geophysical Commission, Moscow

Michell J (1761) Conjectures concerning the cause and observations upon the phenomena of earthquake... of Novemver 1755. Phil. Trans. of the Royal Soc. London 51:566–634

Mouchketow J (1903) Materieux...sur le tremblement de terre d'Alkhalkhalaki du 19. Decembre 1899. Mem du committe Geol., Nouv., Livraison 1, pp 1–80 (in Russian)

Münster S (1544 →1618). Cosmographia Universa, A. Kober, Nüremberg

Noggerath JJ (1847) Das Erdbeben vom 29. Juli 1846 in Rheingebiet und den benachbarten Landern, 60p and 1 map, Bonn, Germany

Pagaczewski J (1972) Catalogue of earthquake in Poland in 1000–1970 years. Publs Inst Geoph Pol Acad Sci 51:3–61 (with 2 maps)
Placanica A (1985) Il filosofo e la catastrofe: un terremoto del Settecento. Einaudi, 1985, 244p (Biblioteca di cultura storica, 155), Torino
Rasch J (1582) Von Erbidem, etliche Tract..., A. Berg, Münich
Reid HF (1914) The Lisbon earthquake of November 1, 1755. Bull Seismol Soc Amer 4:53–80
Réthly A (1952) A Karpátmedeneék foldrengései (445–1918). Akademiai Kiado, Budapest
Réthly A (1960) Report on the earthquake of Moor in the year 1810. Akadémia Kiado, Budapest
Rossi MS, Forel FA (1881) Macroseismic intensity scale description, Jahrbuch d. Tellurischen Observatoriums zu Bern, No. 7, Bern
Rudzki MP (1909) Fizyka Ziemi (physics of the earth). Akad Um Kraków, Wydaw (in Polish)
Rudzki MP (1911) Physik der Erde. Chr. H. Tauch, Leipzig
Sadebeck M (1858) Über die in Schlesien am 15. Januar 1858 wahrgenommene Erdschutterung, Jahrsberichte d. schles. Gessell. f. vaterl. Kultur, Breslau
Schenk V, Schenková Z (2007) Maps of Seismic Zones in Recent Czech National Codes. Exploration Geophysics, Remote Sensing and Enviroment, IV. 2. in: Poodří, čas. obyvatel Horní Odry, 10. roč., Brno, pp 14–18
Schmidt JFJ (1858) Bericht über das Erdbeben am 15. Januar 1858 in Karpathen und Sudeten, Sitzungber. d. k. Acad. d. Wiss. in Wien, math.-naturwiss. Cl., Bd. XXV, Wien
Schmidt P (1980) Gedanken zum Umbruch in der europänischen Seismologie während des 18. Jahrh. Z. geol. Wiss 8(2):189–206
Schweitzer J (undated) German national report, part A. Early German contributions to modern seismology, Chapter 79.24, www.dag/online.de/geschichte/Johannes/Ch79/24CD part A.pdf
Seebach von K (1873) Das Mittedeutsche Erdbeben vom 6. Marz 1872, H. Haissel, Leipzig, pp 1–192
Shröder W (2000) Emil Wiechert – Physiker, Geophysiker, Wissenschaftorganisator, Mitteil. d. Arbeitskreises Geschichte der Geophysik, 19-Jhrg., Heft 1–2, 216p. ISSN: 0179-5658
Sponheuer W (1952) Erdbeben katalog Deutschlands und der angrenzenden Gebiete für die Jahre 1800 bis 1899. Mitt. d. deutsch. Erdbeben Dienstes Nr. 3, hrsg. v. Zentralinst. f. Erdbeben Forschung, Berlin (Akademie), Jena, 195p
Sturm F (1903) Das Sudetische Erdbeben vom 10. Januar 1901. Neuen Jahrbuch f. Mineralogie, Geologie und Paläntologie, Beilage – Band XVI. pp 199–240, with 2 maps, E. Naegele, Breslau
Vaněk J, Kozák J (2007) First macroseismic map with geological background composed by L.H. Jeitteles. Acta Geoph Geod 55:594–606
Varga P (2008) History of early isoseismal maps. Acta Geod Geoph Hung 43:287–307
Vivenzio G (1783) Istoria e teoria de terremoti in particolare di quelli della Calabria e di Messina di 1783, Napoli
Vivenzio G (1788) Historia de terremoti avvenuti nella Provinzia della Calabria ulteriore e nella citta di Messina nel anno 1783, Napoli
Volger GHO (1856) Untersuchungen über das letztjahrige Erdbeben in Central Europa, Mitteilungen aus Justus Perthes Geographischer Anstalt, H. Hässel, Gotha
Volger GHO (1857–1858). Untersuchungen über das Phanommen der Erdbeben. Perthes, Gotha
Woldřich JN (1901a) Zemětřesení v SV Čechách ze dne 10. ledna 1901. Rozpravy Čes. Akad. pro vědy, slovesnost a umění, roč. **10**, třída II., c. 25, pp 1–32, (1 map) (in Czech)
Woldřich JN (1901b) Das Nord Ost Böhmische Erdbeben vom 10. Jänner 1901. Sitzungsb. d. k. Akademie d. Wiss. in Wien., math.–naturwiss. Kl., Bd. 110, Abt.1

Chapter 3
Seismicity in Poland: Updated Seismic Catalog

Barbara Guterch

Abstract A short overview is presented of seismicity in Poland, based on historical data and, since the 1980s, instrumental records. As a level of seismicity is low, in most of the country no earthquakes have been recorded instrumentally, seismicity is extremely poorly recognized. There are areas known historically from the occurrence of earthquakes of $I_0 = 6–7$ but these are single events, never ever repeated there before and afterwards. This concerns mainly the Sudetic Foreland (Paleozoic Platform), the wide Teisseyre-Tornquist Zone and the Precambrian Platform. The most seismically active and the best recognized in Poland are Western Carpathians, along the Pieniny Klippen Belt, especially in the Orawa-Nowy Targ Basin, where continuing tectonic activity has been confirmed by carried out since August, 2008 seismic monitoring. A catalogue of earthquake recorded in Poland in the years 1496–2014, updated and supplemented by foreshocks and aftershocks, is presented.

Keywords Earthquakes · Seismicity · Macroseismic data · Magnitudes and intensities of earthquakes

3.1 Introduction

Poland is a country of low seismicity level and disastrous earthquakes do not occur. According to the present knowledge, macroseismic intensities of the strongest earthquakes historically noted in the last millennium and instrumentally recorded since the 1960s, have never reached 8 in the EMS scale. The magnitudes M_w of the strongest earthquakes in the same period would hardly reach 5.5.

B. Guterch (✉)
Institute of Geophysics, Polish Academy of Sciences, ul. Księcia Janusza 64, 01-452 Warsaw, Poland
e-mail: bguterch@igf.edu.pl

B. Guterch and J. Kozák (eds.), *Studies of Historical Earthquakes in Southern Poland*, GeoPlanet: Earth and Planetary Sciences,
DOI 10.1007/978-3-319-15446-6_3

The seismic recognition is poor. The main problem in seismicity studies, especially in such areas of low seismicity as Poland, is the lack of reliable earthquake catalogues. Data are collected from historical sources of various times and quality. Earthquake parameters for the same events reported in different catalogues often vary. Misprints done by one author can be treated as a new event by another author and it is a reason why events could be multiplied. Moreover, particularly in state border areas, seismic events may be duplicated.

Information on seismic events occurring in Poland, or supposed to occur in Poland, can mainly be found in the following catalogues: Laska (1902), Sieberg (1940), Sponheuer (1952), Kárník et al. (1958), Pagaczewski (1972), Grünthal (1988), Garetsky et al. (1989). The first catalog of earthquakes in Poland, compiled by Laska (1902), professor of astronomy at the University in Lwów, was descriptive. Pagaczewski's (1972) catalogue gives, for some events, the focal parameters: location, depth, intensity in the source in the MSK-64 scale, and magnitude. This catalogue, compiled for all earthquakes observed in Poland in the period 1000–1972, regardless whether they were only macroseismically felt from distant epicenters, or their epicenters were located in Poland, comprises 111 events. Most quakes were only mentioned; presumably, these quakes were weak, with intensities less than 6, and it is only their rarity that captured people's attention.

In 1997–1998, a study was performed by Z. Schenková, V. Schenk and P. Kottnauer for the Czech Republic; by B. Guterch and H. Lewandowska–Marciniak for Poland; and by P. Labak for Slovakia, to develop a common catalogue of these countries—the CZ-PL-SK catalogue (Schenková et al. 1999). All the events in the CZ-PL-SK catalogue have been analyzed by looking for the source data and searching over regional catalogues. This review eliminated errors in data and location that caused earthquake duplications. The seismicity data file was made homogeneous with respect to the epicenter intensities for the majority of events were historical earthquakes. Intensities in the CZ-PL-SK catalogue were estimated in the MSK-64 scale.

Homogeneous regional catalogues were the basis for computing the European and global peak ground acceleration hazard maps, in the framework of Global Seismic Hazard Assessment Program (GSHAP), by Grünthal et al. (1999), Giardini et al. (1999). The CZ-PL-SK catalogue together with national catalogues of Central Europe gave the input data for calculation of the seismic hazard in terms of macroseismic intensities and peak ground attenuation (PGA) for the Czech Republic, Poland, and Slovakia (Schenk et al. 2000, 2001).

National catalogues have being currently updated. The catalogue of earthquakes in Poland was published by Guterch and Lewandowska-Marciniak (2002) and later in Polish by Guterch (2009a). Catalogue of earthquakes in Poland presented here is based, with some updating, on the CZ-PL-SK catalogue. It was supplemented by aftershocks and foreshocks and covers the years 1496–2014.

As opposed to the low tectonic seismicity level, mining induced seismicity in Poland is considerably high, one of the highest in the world, exceeding $M = 4$ almost every year. There are three main regions of induced seismicity in Poland: Upper Silesia, where coal has been mined since XIX century; Lubin Copper

Mining Basin, where exploitation started in 1968; and Bełchatów in central Poland, where brown coal has been mined from the surface since year 1976. Seismic monitoring is carried out in all mines exposed to induced seismicity.

3.2 Review of Records About Seismic Events in Poland Until the XV Century

Mentions about earthquakes felt in Poland until the XV century can be found in chronicles by Jan Długosz (1415–1480 a chronicler and diplomat). Later on, great Polish scholars: Maciej from Miechów called Miechowita (1457–1523, a historian, chronicler, geographer), and Maciej Kromer (1512–1589, a historian, cartographer, diplomat) repeated (see the Refs.: Miechowita 1986; Kromer 1611) information about earthquakes after Długosz and did not mention any new events.

Długosz's notes are short and enigmatic. There is no base to estimate earthquake parameters for the event noted for the year 1016 by Długosz (quoted from the edition of 1873, as Długosz 1873, **X**, p. 210): '*Sub huis autem anni tempore tremor terrae Poloniae Regno accidens prodigii loco habitus est' [The earthquake which occurred this year in the Polish kingdom was regarded as a miracle]*. Later information repeated the same note after Długosz but Jeitteles (1860) who added "*Erdbeben zu Krakau*". This Jeitteles's supplement is not supported by source reference.

Długosz's record for the year 1200 (quoted from the edition of 1873, as Długosz 1873, **XI**, p. 158) "*Quinta Maii mensis die terrae motus in Polonorum regione, vicinisque Regnis et terris sub meridiano tempore exertus et frequenti dierum spatio resemptus, plerasque turres, domos et munitiones attrivit. Quae res raro in Polonia contigens, velut prodigiosa et apud plerosque in religionem versa vedibatur" [On May 5, an earthquake in Poland and adjacent countries occurred at high noon, and was several times repeated in next days, turning down turrets of houses and towns. Such phenomena, very rarely occurring in the Polish country, were regarded as a great peculiarity, and overwhelmed some people with excessive panic*[1]*]* probably concerns the Alpine earthquake on May 4, 1201, of $I_0 = 9$ and testifies that Poland was in macroseismic area of the main earthquake and its aftershocks.

Długosz' record for the year 1257 (quoted from the edition of 1873, as Długosz 1873, **XI**, p. 365) is the most interesting mention about Kraków: '*Pridie Calendas Februarii Cracoviae et in universis Polonorum urbibus et regionibus grandis terrae motus, apud Polonos rei huiuscemodi insuetos, pro monstro et religione habitus, editus est Tertiarum hora' [At Tertiarum hora on January 31, in Kraków and all towns and regions of Poland, a great earthquake occurred, to evoke an astonishment for the inhabitants quite unaccustomed to such events]*. After Bohemian

[1] Translated from Polish version by Anna Dziembowska.

and Silesian sources it seems that it concerns the year 1259. Although it is supposed that this earthquake could be the strongest ever recorded in area of Poland, its epicenter parameters are impossible to determine (Guterch et al. 2015a, this issue).

All positions in Pagaczewski's catalogue between the years 1433 and 1443 are likely to concern the June 5, 1443, Žilina earthquake of $I_0 = 8$, in Slovakia (Pieniny Klippen Belt) or its aftershocks. Towns in Poland: Kraków, Wrocław and Brzeg, suffered serious damage caused by this earthquake (Rocznik Magistratu Wrocławskiego 1878, p. 686; Długosz 1877, **XIII**, p. 691, Grünhagen 1870). After Długosz, the vaulting of Saint Catherina church fell down on this day.

Catalogs of Koláček (1921), Kárník et al. (1958), Pagaczewski (1972), Guterch and Lewandowska-Marciniak (2002) report an earthquake in 1483 in Brzeg. However, in chronicles in Brzeg, a confirmation of such an earthquake has not been found. After Grünhagen (1870), the event that happened in Brzeg was dated on June 5, 1443; it was: "*Erdbeben, in Folge dessen ein Stück Gewölbe von der Nikolaikirche einstürtz*" [*an earthquake, as a result of which a piece of wall from the (St.) Nicholas church fell down*]; nothing like this is mentioned for the year 1483. Thus, an event in 1483 would be a result of misprint probably done by Koláček (1921).

Similarly, in archives of Toruń (Archiwum Państwowe, Toruń) no information has been found about such a spectacular event in XVI century as an earthquake in January 1572 (Pagaczewski 1972; Laska 1902; Guterch and Lewandowska–Marciniak 2002), while heavy storms have been listed, i.e., in Zerneke (1711). Thus, the seismic event in January 1572 seems to be questionable.

3.3 Catalogue of Earthquakes in Poland

There are only a few earthquakes listed in the catalogue in the XV–XVII centuries for which epicentral parameters could be estimated. It is supposed that these are extreme cases, as only such cases could render people's attention, weaker than $I_0 < 6$ being only occasionally recorded.

The list presented in Table 3.1 includes only the earthquakes occurring in Poland, without the events felt in Poland from epicenters located in other regions, i.e., the Alps, or the Vrancea and Zilina regions in the Carpathians. Epicentral data, i.e., dates, time, coordinates, and intensities have been given together with their accuracies. Intensities have been assessed in the EMS-98 scale. Coordinates of seismic events since the year 1989 have been determined instrumentally, but the one on June 1, 1994, recorded only macroseismically and 2 events on October 13, 1995, recorded by 1 seismic station. Macroseismic coordinates of epicenters, since 1989, have been given in remarks for events where a sufficient number of intensity data points were available. Earthquake epicenters in Poland in the intensity EMS scale are presented in Fig. 3.1. Data of seismic events in the Czech Republic and Slovakia for $I_0 \geq 4$ are after Schenková et al. (1999).

Table 3.1 Catalogue of earthquakes in Poland

No.	SQ	TE	Date	DA	Time h m s	Coordinates		CA	I_o	I_0A	h	Magnitude				Reg.	Remarks	References
												Macroseismic		Instrumental				
						φ	λ					M_L	M_W	M_L	M_W			
1	3		1259 01 31	4					>7	3								KMM58; PA72; DLUG; GLMMa
2	3		1496 06 23	5		50.3	17.3	5	5	3		3.5	3.3			III		SIE40; KMM58, PA72
3	3		1562 02 10	5		50.5	16.7	5	7	2		5.2	4.9			III		LV96; KO16; KMM58; PA72
4	3		1594 09 15	3		51.1	15.9	5	5	3		3.5	3.3			III		KMM58; PA72
5	2		1606	5		53.2	16.2	3	6	2		4.3	4.0			I		LA02; PA72
6	3		1615 02 13	5		50.5	16.7	4	5	3		3.5	3.3			III		LV96; KMM58; PA72
7	3		1716	5		49.4	19.9	4	6	3		4.3	4.0			VI		JE60; LA02; KMM58; PA72; GU06
8	3		1717 03 11	5		49.4	19.9	4	7	3		5.2	4.9			VI		JE60; LA02; KMM58; PA72; GU06
9	3		1751 06 31	1		50.8	15.6	4	6–7	2		≥4.3	≥4.0			III		LV96; KMM58; PA72
10	2		1774 01 26/27	5		50.1	18.2	4	7	2		5.3	4.9					JE60; SIE40; KMM58; PA72
11	2	F	1775 01 24	1	02 05											III		Zeplichal (1775); JE60; SIE40; KMM58; PA72
		M	1775 01 24		02 09	51.1	17.0	5	4	3		2.7	2.6					
		A	1775 01 24		02 12													
12	3		1778 05 10	1		51.0	15.9	4	5–6	3		≥3.5	≥3.3			III		AST (1778); JE60; PA72
13	2		1785 08 22	1	07	49.9	18.8	3	6	1	16	4.6	4.3			IV	Macros. map	KMM58; OL65; GLMMa; and others

(continued)

Table 3.1 (continued)

No.	SQ	TE	Date	DA	Time h m s	Coordinates		CA	I_o	I_0A	h	Magnitude				Reg.	Remarks	References
												Macroseismic		Instrumental				
						φ	λ					M_L	M_W	M_L	M_W			
14	2		1786 01 03	5		53.4	14.6	6	3–4	2		≥1.9	≥1.9			I		LA02; GR88; PA72
15	3	F	1786 02 26	1	20				4	3						IV		KMM58; OL66b; PA72; GLMMa; and others
	3	F	1786 02 26		22				4	3								
	2	F	1786 02 27		01	50.0	18.0	4	5–6	2								
	2	M	1786 02 27		04	49.8	18.3	3	7	1	35	5.6	5.3				Macros. map	
	3	A	1786 02 27		18				4	3								
16	3		1786 03 04	1		50.5	16.7	4	4	3		2.7				III		SIE40; KMM58; OL66b; PA72
17	3		1786 10 03	1		49.7	18.6	4	4	3		2.7				IV		RE25; KMM58; OL66a
18	2	M	1786 12 03	1	17	50.4	19.3	4	7	1	35	5.6	5.3			IV	Macros. map	KMM58; OL66a; PA72; GLMMa, GLMMb; and others
	3	A	1786 12 04		03				3	2								
	3	A	1786 12 09		09 08				4	2								
19	3		1790 03 13	1		51.1	17.0	5	4	3		2.7	2.6			III		HO24; HO41; SIE40; KMM58; OL66a; PA72
20	3		1799 02	1		51.1	17.0	5	4	3		2.7	2.6			III		HO24, JE60, KMM58
21	3		1799 10	3	20	50.9	15.75	4	4	2		2.7	2.6			III		HO24, JE60, PA72, KMM58
22	3	F	1803 01 08	1	13 55				4–5	2		≥2.6	≥2.6			II		LA02; PA72; GA89
	3	M	1803 01 08		23 15	53.1	23.1	4	5–6			≥3.5	≥3.3					
23	3		1823 09 09	1		50.2	17.8	4	4	2		2.7	2.6			III		KMM58, PA72
24	3		1824 12 13/14	1		51.8	17.6		4–5	2		≥2.7	≥2.6			I		Turno (1824); LI13
25	2		1829 06 02/03	1		50.7	15.7	4	5	2		3.5	3.3			III	3 shocks	LV96; SP52; KMM58; PA72

(continued)

Table 3.1 (continued)

No.	SQ	TE	Date	DA	Time h m s	Coordinates		CA	I_o	I_0A	h	Magnitude				Reg.	Remarks	References
												Macroseismic		Instrumental				
						φ	λ					M_L	M_W	M_L	M_W			
26	2		1837 02 05	1	23	50.68	21.04	3	6	2		4.3	4.0			I		TZKP manuscript
27	3		1837 02 18	1	06	50.2	19.2	3	5	2		3.5	3.3			IV		TZKP manuscript
28	3		1841 08 06	1	14	51.5	20.6	3	5–6	2		≥3.5	≥3.3			I		TZKP manuscript
29	3		1842 02 24	1	14 07	50.0	20.0	5	4	3		2.7	2.6			V		LA02; PA72
			1842 02 24		20 27													
30	3		1842 03 08	5		50.0	20.0	5	4	3		2.7	2.6			V		SP52; KMM58; PA72
31	3		1855 01 15	1	08 30	49.8	18.8	4	4	3		2.7	2.6			IV		LA02; SP52; KMM58; PA72
			1855 01 15		09 45													
32	3		1857 12 16/17	1		49.7	20.7	4	4	3		2.7	2.6			V	A few shocks	LA02; PA72
33	3		1872 12 26	1		49.8	19.0	4	4	3		2.7	2.6			IV		LV96; KMM; PA72
34	2		1876 07 12	1		49.8	18.6		4			2.7	2.6			IV		POH97; KMM58; PA72
35	2		1877 11 25	1	04	50.4	17.0	4	5	2		3.5	3.3			III		LV96; KMM58; PA72
36	2	F	1895 06 10	1	11				3	3						III		Leonhard and Voltz (1895); LV96, Dathe (1897); SP52; KMM58; OL62; PA72; Kozák (2015), this issue
	2	F	1895 06 11						3	3								
	2	F	1895 06 11		04				3	3								
	2	F	1895 06 11		08 30				3	3								
	2	M	1895 06 11		09 27	50.75	17.0	3	6–7	1	8	≥4.4	≥4.1				Macros. maps	
	2	A	1895 06 11		10				3	3		2.4						
	2	A	1895 06 11		18				3	3		2.4						
37	3		1905 07 19	5		53.4	14.6	4	3–4	2		≥1.9	≥1.9			I		LE13
38	3		1907 08 29			54.0	14.6	6	4	4		2.7	2.6			I		GR88

(continued)

Table 3.1 (continued)

No.	SQ	TE	Date	DA	Time h m s	Coordinates		CA	I_o	I_0A	h	Magnitude				Reg.	Remarks	References
												Macroseismic		Instrumental				
						φ	λ					M_L	M_W	M_L	M_W			
39	3	S	1908 12 30	1	03 35	54.3	22.3	4	4	2		2.7	2.6			II		Tornquist (1909); OL62
40	2		1909 02 11	1	~23	54.1	15.6	3	6	2		4.3	4.0			I		LE13; OL62; PA72
			1909 02 12		~06													
41	3		1909 05 06	1		49.4	21.0	4	4	2		2.7	2.6			V		OL62; PA72
42	3		1912 12 01	9		54.7	17.5	4	4	2		2.7	2.6			I		OL62; PA72
43	3		1920 09 13	9	23	53.6	15.2	9	4	9		2.7	2.6			I		GR88
44	3		1924 12 15	9	22 45	53.9	14.3	9	4	9		2.7	2.6			I		GR88
45	2	M	1926 11 20	1		51.28	19.90	3	6	2		4.3	4.0			I		Fleszarowa (1926)
	2	A	1926 11 22						4	2		2.7	2.6					
46	2	F	1928 06 03	1	01	54.2	16.0	3	4	2		2.7	2.6			I		Bülow (1928); OL62; PA72; GR88
	2	M	1928 06 11		15				5	2		3.5	3.3					
47	2	S	1932 02 06	1	04	50.6	21.2	3	5–6	2		≥3.5	≥3.3			I		JA33; PA72; OL62
48	2	S	1932 02 06	1	04 15	51.75	22.20	3	6	2		4.3	4.0			I		JA33; PA72; OL62
49	2	S	1932 02 06	1	04 15	50.61	20.32	3	5–6	2		≥3.5	≥3.3			I		JA33; PA72; OL62
50	2	S	1932 02 09	1	04	50.6	21.2	3	5–6	2		≥3.5	≥3.3			I		JA33; PA72; OL62
51	2	S	1932 02 10	2		50.60	20.03	2	6	1		4.3	4.0			I		JA33; PA72; OL62
52	2	S	1932 02 11	1	22	50.6	21.2	3	5–6	2		≥3.5	≥3.3			I		JA33; PA72; OL62
53	2	S	1932 02 11	1	22 30	51.75	22.20	3	6	2		4.3	4.0			I		JA33; PA72; OL62
54	2	S	1932 02 11	1	23	50.85	20.30	3	5–6	2		≥3.5	≥3.3			I		JA33; PA72; OL62

(continued)

Table 3.1 (continued)

No.	SQ	TE	Date	DA	Time h m s	Coordinates φ	Coordinates λ	CA	I_o	I_0A	h	Magnitude Macroseismic M_L	Magnitude Macroseismic M_W	Magnitude Instrumental M_L	Magnitude Instrumental M_W	Reg.	Remarks	References
55	2	S	1932 02 11	1	23	50.87	20.70	3	5–6	2		≥3.5	≥3.3			I		JA33; PA72; OL62
56	2	S	1932 02 11	1	23	50.61	20.32	3	5–6	2		≥3.5	≥3.3			I		JA33; PA72; OL62
57	2	S	1932 02 11	1	23	51.20	22.65	3	5–6	2		≥3.5	≥3.3			I		JA33; PA72; OL62
58	2	S	1932 02 24	1	19	52.60	20.05	2	6	1		4.3	4.0			I		JA33; PA72; OL62
59	2	F	1935 03 23	1	08 12				4–5	2		≥2.6	≥2.5			VI		OL62; PA72
		M	1935 03 23		23 46	49.45	19.85	3	7	1	~3	5.0	4.7					
		A	1935 03 23		05 29				5–6	2		≥3.4	≥3.2					
60	3		1942 03	5		49.4	19.9	4	4–5	3		≥2.7	≥2.6			VI		Author's data
61	1	F	1966 03 10		22 37 23				2–3	2						VI		Mazur (1968); PA72
		F	1966 03 14		12 59 41				3	2		1.8	1.8					
		F	1966 03 17		00 41 53				3–4	2		≥1.8	≥1.8					
		M	1966 03 17		00 53 55	49.3	19.9	3	4–5	1	~3	≥2.6	≥2.5					
		A	1966 03 19		03 01 54				3	2		1.8	1.8					
62	1		1989 10 24		18 04 22	49.32 ± 0.081	19.76 ± 0.075							3.1	2.8	VI		GU05; Seism. Bull. 1989–1990
63	1		1989 11 15		03 26 26	49.33 ± 0.070	19.74 ± 0.078							2.7	2.6	VI		GU05; Seism. Bull. 1989–1990
64	1	F	1992 06 28		23 18 30	49.40 ± 0.108	21.03 ± 0.070							3.2	3.0	V		DE97; GU00; Seism. Bull. 1991–1992
		M	1992 06 29		00 29 48.0	49.50 ± 0.053	21.12 ± 0.037		5–6	1	3	≥3.4	≥3.2	3.6	3.1		49.42°N 21.10°E Macros. map	
		A	1992 06 29		00 34 56.5	49.53 ± 0.080	21.08 ± 0.060		4	1		2.6	2.5	3.2	2.9			
		A	1992 06 29		06 04 50	49.40 ± 0.145	20.95 ± 0.080							2.9	2.5			
		A	1992 06 30		23 34 34	49.33 ± 0.100	21.02 ± 0.050							3.1	2.8			
		A	1992 09 05		20 33 56	49.42 ± 0.162	21.02 ± 0.234							3.3	2.9			

(continued)

Table 3.1 (continued)

No.	SQ	TE	Date	DA	Time h m s	Coordinates		CA	I_o	I_0A	h	Magnitude				Reg.	Remarks	References
												Macroseismic		Instrumental				
						φ	λ					M_L	M_W	M_L	M_W			
65	1	M	1993 03 01		07 42 39.0	49.48 ± 0.067	21.07 ± 0.042		7	2	3	5.0	4.7	4.0	3.6	V	49.45°N 21.10°E Macros. map	DE97; GU00; Seism. Bull. 1993–1994
		A	1993 03 01		08 22 28	49.47 ± 0.064	20.97 ± 0.054		3									
66	3		1994 06 01	1	15	53.85	22.90	4	5–6	2		≥3.5	≥3.3			II		Author's data
67	1		1994 11 09		20 10 25	49.74 ± 0.260	18.62 ± 0.304								2.6	IV	49.73°N 18.38°E (GFU)	Seism. Bull 1993–1994; http://www.czechgeo.cz/gfu-bulletin/?do=process
68	1	M	1995 09 11		04 02 19	49.39 ± 0.072	19.90 ± 0.174		5	1	2	3.7	3.4		3.3	VI	Macros. map of the main shock: 49.40°N 19.86°E	GU05; Seism. Bull. 1995
		A	1995 09 11		05 02 22	49.39 ± 0.073	19.88 ± 0.176								2.7			
		A	1995 09 11		09 02 34	49.38 ± 077	19.84 ± 0.180								2.6			
		A	1995 09 11		17 17 20	49.39 ± 0.078	19.88 ± 0.188								2.8			
69	1		1995 10 13		05 47 16	49.40	20.10	1	4	1	2	2.6	2.5			VI	Macros. map	GU05; Seism. Bull. 1995
			1995 10 13		06 02 33	49.41	20.10	1	4	1	2	2.6	2.5				Macros. map	
70	1	M	2001 07 07		10 10 27	49.5	20.6	2							2.8	V	+10 aftershocks of 1.9–2.4 M	Seism. Bull. 2001 (2003)
71	1		2003 03 12		13 20 27	49.4	19.8	2							2.6	VI		Seism. Bull. 2003 (2005)

(continued)

Table 3.1 (continued)

No.	SQ	TE	Date	DA	Time h m s	Coordinates		CA	I_o	I_0A	h	Magnitude				Reg.	Remarks	References
												Macroseismic		Instrumental				
						φ	λ					M_L	M_W	M_L	M_W			
72	1	M	2004 11 30		17 18 38.0	49.445 ± 0.031	19.889 ± 0.033		7	1	3	5.0	4.8		4.5	VI	49.40°N 19.93°E Macros. map	Guterch (2006); Seism. Bull. 2004; Seism. Bull. 2005; Wiejacz and Debski (2009)
		A	2004 12 01		23 25 15.6	49.466 ± 0.103	19.967 ± 0.132		4	1		2.6	2.5		2.8		49.40°N 19.92°E Macros. map	
		A	2004 12 01		23 50 13.0	49.32 ± 0.069	19.91 ± 0.091		4	1		2.6	2.5		2.7			
		A	2004 12 02		18 25 39.2	49.395 ± 0.065	19.843 ± 0.069		5	1	3	3.4	3.2		3.6		49.39°N 19.88°E Macros. map	
		A	2004 12 03		10 07 02.9	49.34 ± 0.056	19.86 ± 0.074								2.7			
		A	2004 12 09		01 09 05.8	49.378 ± 0.051	19.884 ± 0.065		5	1	~3	3.4	3.2		3.3		49.40°N 19.93°E Macros. map	
		A	2004 12 13		00 05 31.8	49.426 ± 0 .069	19.874 ± 0.077								2.7			
		A	2005 01 23		23 33 20.0	49.395 ± 0.067	19.930 ± 0.063		4-5	1	~3	≥2.6	≥2.5		3.1		49.40°N 19.92°E Macros. map	
		A	2005 01 29		17 16 56.2	49.412 ± 0.059	19.931 ± 0.061		5	1	~3	3.4	3.2		3.4		49.40°N 19.93°E Macros. map	
		A	2005 02 07		06 08 42.3	49.447 ± 0.139	19.967 ± 0.163.		3-4	1	3-5	≥2.2	≥2.0		2.7			
		A	2005 02 18		01 32 48.8	49.37 ± 0.080	19.78 ± 0.091								2.7			
		A	2005 06 02		07 43 27.6	49.412 ± 0.059	19.858 ± 0.074		5	1		3.4	3.2		3.2		49.40°N 19.93°E Macros. map	
		A	2005 08 24		15 46 12.4	49.395 ± 0.091	19.936 ± 0.135								2.8		+ ~290 aftershocks of 1.5–2.6 M_w during a year, about 40 % within 24 h after the main event	

(continued)

Table 3.1 (continued)

No.	SQ	TE	Date	DA	Time h m s	Coordinates		CA	I_o	I_0A	h	Magnitude				Reg.	Remarks	References
												Macroseismic		Instrumental				
						φ	λ					M_L	M_W	M_L	M_W			
73	1		2006 06 25		01 12 26.1	49.26 ± 0.061	20.01 ± .087								3.2	VI		Seism. Bull. 2006
74	1		2007 05 06		07 32 31	52.005	17.490							2.8	2.9	I	52.00°N, 17.46°E (GFU)	Trojanowski et al. (2012); http://www.czechgeo.cz/gfu-bulletin/?do=process
75	1	M	2007 09 25		21 32 13.2	49.588 ± 0.004	20.914 ± 0.013								2.6	V	+10 fore-shocks and 2 aftershocks	Wiejacz et al. (2008)
76	1		2011 11 22		14 55 15.04	49.419	20.258				4.2			2.1	2.8	VI	A swarm type series + ~55 tremors of 1.5–2.6 Mw, since November 01 until December 31, 2011	Trojanowski et al. (2015); http://private.igf.edu.pl/~nlkozlak/
77	1		2012 01 06		15 37 56	52.007 ± 0.014	17.533 ± 0.029		5	1	4.5	3.5		3.8	3.5	I	Macros. map	LI13

(continued)

Table 3.1 (continued)

No.	SQ	TE	Date	DA	Time h m s	Coordinates φ	Coordinates λ	CA	I_o	I_0A	h	Magnitude Macroseismic M_L	Magnitude Macroseismic M_W	Magnitude Instrumental M_L	Magnitude Instrumental M_W	Reg.	Remarks	References
78	1	S	2013 02 17		13 37 14.5	49.425	20.265		Felt					2.7	3.1	VI	The whole event 78: A swarm-type series ~110 tremors of 1.5-2.6 M_W February 16-March 31, 2013; Macros. map of the main shock: 49.45°N, 20.23°E (intensity data points)	http://private.igf.edu.pl/~nlkozlak/; IGFPAN; Author's data
		S	2013 02 23		21 41 05.7	49.436	20.260		Felt					2.0	2.8			
		M	2013 03 01		21 23 12.9	49.432	20.266		4–5	1		≥2.7		3.3	3.5			
		S	2013 03 02		10 50 13.5	49.432	20.265		Felt					2.1	2.7			
		S	2013 03 02		10 50 35.8	49.427	20.266		3					3.0	3.1			
		S	2013 03 02		10 51 42.1	49.414	20.269		Felt					2.1	2.7			
		S	2013 03 02		17 13 04.2	49.422	20.267		Felt					2.2	2.6			
		S	2013 03 04		00 05 11.9	49.420	20.267							2.1	2.7			
		S	2013 03 25		01 06 28.4	49.418	20.266							2.7	3.2			
		S	2013 04 20		23 26 04.4	49.415	20.271							2.2	2.7			
		S	2013 04 21		06 57 51.7	49.432	20.264							2.7	3.1			
79	1	M	2013 08 08		10 54 06.8	49.388	20.251								3.0	VI		http://private.igf.edu.pl/~nlkozlak/; IGFPAN

(continued)

Table 3.1 (continued)

No.	SQ	TE	Date	DA	Time h m s	Coordinates		CA	I_o	I_0A	h	Magnitude				Reg.	Remarks	References
												Macroseismic		Instrumental				
						φ	λ					M_L	M_W	M_L	M_W			
80	1	M	2014 04 01		09 33 36.3	49.435	20.475		Felt		4.9			2.8	3.1	V	+3 aftershocks	Plesiewicz and Wiszniowski (2015), this issue; http://private.igf.edu.pl/~nlkozlak/

Updated December 2014
SQ source quality: *1* instrumental data, *2* reliable sources data, *3* uncertain sources data
TE type of event: *M* main, *F* foreshock, *A* aftershock, *S* swarm
DA date accuracy: *1* well assessed, *2* uncertain day, *3* uncertain month, *4* uncertain year, *5* impossible to assess
CA coordinates accuracy: *1* $(\Delta\varphi, \Delta\lambda) < 0.05°$, *2* $0.05° < (\Delta\varphi, \Delta\lambda) < 0.1°$, *3* $0.1° < (\Delta\varphi, \Delta\lambda) < 0.25°$, *4* $0.25° < (\Delta\varphi, \Delta\lambda) < 0.5°$, *5* $(\Delta\varphi, \Delta\lambda) > 0.5°$, *6* impossible to access
I_o—intensity in the EMS-98 scale
I_oA—intensity accuracy: *1* well assessed, *2* $\Delta I_o = \pm 1$, *3* $\Delta I_o > 1$, *4* impossible to assess
M_L—local magnitude
M_w—seismic moment magnitude
Reg—region
Magnitudes derived from macroseismic data—after Grünthal et al. (2009), i.e., relations [12] and [2]; default focal depth of 5 km, if the depth is unknown
Abbreviations to references: BU28 = Bülow (1928); DE97 = De̜bski et al. (1997); DLUG = Długosz (1877); GA89 = Garetsky et al. (1989); GLMMa = Guterch et al. (2015a, this issue); GLMMb = Guterch et al. (2015b), this issue); GR88 = Grünthal (1988); GU00 = Guterch et al. (2000) GU05 = Guterch et al. (2005); GU06 = Guterch (2006); GFU = Geofyzikální ústav Akademie veˇd ČR; HO24 = Hoff (1824); HO41 = Hoff (1841), IGFPAN = Database of the Department of Seismology, Institute of Geophysics, Polish Academy of Sciences; JA33 = Janczewski (1933); JE60 = Jeitteles (1860); KMM58 = Karnik et al. (1958); KO16 = Koláčˇek (1916); LA02 = Laska (1902); LE13 = Lehman (1913); LI13 = Lizurek et al. (2013); LV96 = Leonhard and Voltz (1896); OL62 = Olczak (1962); OL65 = Olczak (1965); OL66a = Olczak (1966a); OL66b = Olczak (1966b); PA72 = Pagaczewski (1972); POH97 = Pohludka (1897); Seism. Bull. 1989–1990 = Draber et al. eds. (1998); Seism. Bull. 1991–1992 = Draber et al. eds. (2000a); Seism. Bull. 1993–1994 = Draber et al. eds. (2000b); Seism. Bull. 2001 = Draber et al. eds. (2003); Seism. Bull. 2003 = Zuchniak et al. eds. (2005); Seism. Bull. 2004 = Guterch ed. (2007); Seism. Bull. 2005 = Guterch ed. (2008); Seism. Bull. 2006 = Guterch ed. (2009a, b); SIE40 = Sieberg (1940); SP52 = Sponheuer (1952). Other abbreviations and details in the section References

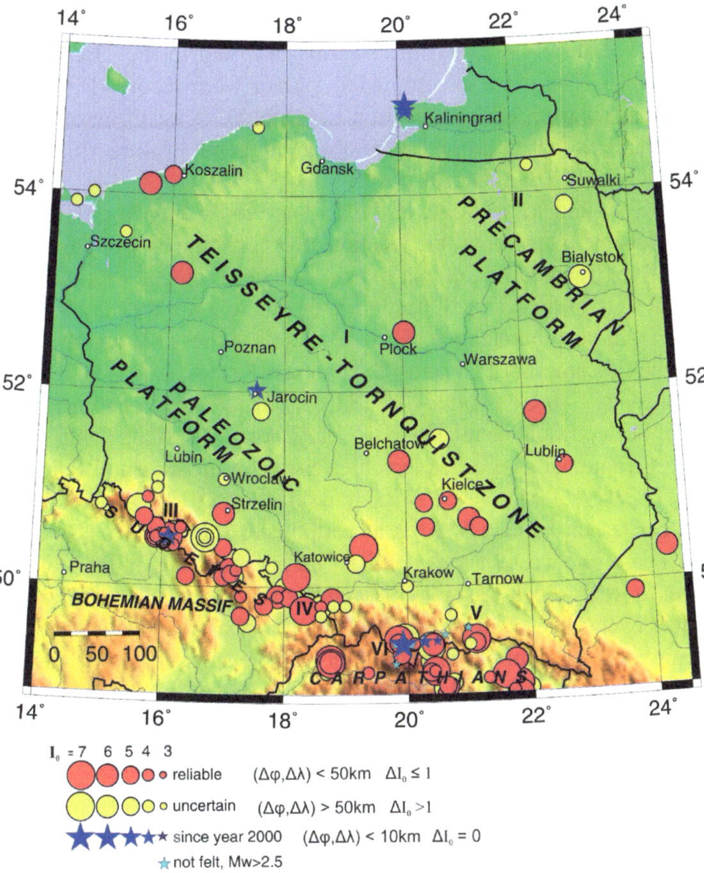

Fig. 3.1 Earthquake epicenters in Poland since the XVI century in EMS intensity scale. Location of earthquakes in Slovakia and the Czech Republic in border areas for $I_0 \geq 4$ after Schenková et al. (1999). *I, II, III, IV, V, VI*—regions of earthquake occurrence

3.4 Regions of the Occurrence of Seismic Events in Poland

Epicenters of seismic events in Poland are located mostly in the Western Carpathians, the Poland-Slovakia border areas; in the Sudetes, the Poland-Czech Republic border areas; and along the broad Teisseyre-Tornquist Zone (TTZ), being part of the Trans European Suture Zone (TESZ).

3.4.1 Region I: Teisseyre-Tornquist Zone

Seismotectonic origin of earthquakes in the TTZ has not been recognized. The earthquakes are recorded along the broad area of the TTZ, mainly in the West

Pomerania (west of Koszalin), and in the vicinicity of the Holy Cross Mts (south of Kielce). In February 1932, seismic swarms occurred in the TEZ with intensities up to 6 EMS. The main events of the swarm were recorded in a few areas up to 200 km away from each other. In all epicentral areas of the main events of the swarm, cracks in frozen ground were observed, some of them still visible after 4 months (Janczewski 1933; Olczak 1962).

The macroseismic data of events in the TTZ are too scanty to enable a compilation of macroseismic maps. Thus, the location error for some events may even exceed 50 km. Therefore, it would be difficult to associate their sources with the known tectonic faults. In spite of the adopted location errors, there is a correlation of earthquake foci with the main deep boundaries of crustal blocks in the TTZ, mainly along western edge of the Precambrian Platform (Guterch and Lewandowska-Marciniak 2002). Events, of intensity 5–6 on February 6, 1837, around Raków, and on February 11, 1932 south of Kielce, could be related to the Holy Cross Mts. Main Fault.

The first instrumentally recorded tectonic seismic event in central Poland was the one of May 6, 2007 of $M_W = 2.9$ (Trojanowski et al. 2012; Geofyzikální ústav[2]). About 5 years later, on January 6, 2012, an earthquake of $M_W = 3.5$, $I_o = 5$, occurred in the same area of the western edges of the TTZ in the vicinity of the Dolsk fault. In the same area, a weak shock was felt in December 1824 (Lizurek et al. 2013).

All known events recorded in the TTZ share a common feature of very small macroseismic area, that implies shallow foci of the events. Depth of the last event on January 6, 2012, determined after macroseismic data, was also shallow, i.e., 4.5 km (after Lizurek et al. 2013).

3.4.2 Region II: Precambrian Platform

The intra-plate events in the Precambrian Platform are scanty, the region is considered to be of very low seismicity (Garetsky et al. 1989). Within the Polish border there are known 3 events. Accuracy of focal parameter assessment of these event is small, i.e., $(\Delta\varphi, \Delta\lambda) \geq 0.5°$, $\Delta I_o \geq 1$.

A surprise for European seismology was the onset of earthquakes on September 21, 2004 in the Kaliningrad region near the Baltic coast: of M = 5.0 (11:05 GMT) and M = 5.2 (13:32 GMT), (Gregersen et al. 2007, Wiejacz 2006). On the seismic regionalization maps for the year 1998, this region of the Precambrian Platform is situated in the zone where I = 5 should not be exceeded over the 500–5,000 years (Nikonov 2005). Because of such a very low seismic hazard assessment, there were no seismological stations in the area.

In Poland, these earthquakes caused slight damage to individual buildings and dwelling houses in 115 localities, mainly in north and northeastern Poland. Most of damages were non-structural, of Grade 1, very seldom moderate of Grade 2 in the EMS scale, and sustained buildings of vulnerability class A–B. Suwałki

[2] Geofyzikální ústav Akademie věd ČR (GFU) http://www.czechgeo.cz/gfu-bulletin/?do=process.

town, at a distance of about 210 km from the epicenters, suffered most damage in Poland. The local authorities were informed about damage to 45 buildings or dwelling houses. The majority of damages in Suwałki were slight, non-structural to masonry buildings of rather poor technical state in the old town. Damage of Grade 1 was observed also in some few-storied buildings of reinforced concrete and spreading construction (vulnerability B–C) in new settlements at the northern part of Suwałki (Guterch 2009a).

After the 2004 earthquakes, a critical verification was undertaken of historical documents relating to extreme natural events, such as floods on the Baltic coast areas, to find possible associations to an under-sea event, but the results are still inconclusive. Seismic events of $M_L < 2.9$ recorded in the Baltic Sea in the Gulf of Gdańsk in the years 1996-1997 were most likely man-made explosions performed in the Russisn territorial waters (Wiejacz and Dębski 2001).

3.4.3 Region III: Sudetes and Their Foreland

The seismicity of the Sudetes bordered on the SW by the Bohemian Massif, is the strongest in the Hronov-Poříčí fault zone in the Czech Republic. The strongest known earthquake occurred in 1901, January 10 with $I_o = 7$. Macroseismic maps were compiled immediately after the earthquake by Woldřich (1901), Credner (1901), Gränzer (1901), and Sturm (1903) (see also Kozák and Plešinger 2003, Kozák 2015, this issue). On October 25, 2005, an earthquake of $M_L = 3.3$, $I_o = 5$ occurred in the area of the Hronov-Poříčí fault. Two years of seismic observations in the vicinity of the Hronov-Poříčí Fault revealed relatively frequent weak seismic activity (Málek et all. 2008). Continuing tectonic activity of the Fault is confirmed by an evidence of week shocks recorded there every 1 or 2 years (i.e., Zednik et al. 2001).

Other Western Sudetes main areas where seismic events have been recorded are: Karkonosze Mts (Poland–Czech border) and Kłodzko Basin. Epicentral parameters of historically known earthquakes recorded at Kłodzko are of small accuracy, even for the strongest one of $I_o = 7$ on February 10, 1562.

The main areas of earthquakes occurrence in the Eastern Sudetes are Jeseníky Mts. (the strongest known on 24 July 1935 of $I = 5.5$ MSK), Śnieżnik Massif at the Poland–Slovakia border related to the Ramzowa Overthrust; and Opava fault zone in the Moravian Silesian Basin (the strongest known on April 12, 1931, of $I_o = 6$ MSK with a series of foreshocks and aftershocks between March 30 and May 25, 1931) (Kárník et al. 1958).

Sudetic Foreland (Paleozoic Platform) is very weakly recognized seismically. The epicenter data of a few historically recorded seismic events are of small accuracy but the one on June 11, 1895, of $I_o = 6$–7 recorded south of Strzelin. This seismic event occurred only 2 months after the strong Ljubljana earthquake of March 14, 1895. On April 25, 1895, the special Commision (Erdbeben–Commision) was established by the Department of Mathematical-Natural Sciences of the Imperial Academy of Sciences in Vienna to organize systematic macroseismic observations and central collection of reports (Kozák and Plešinger 2003). Thus, this earthquake, similarly to the previously mentioned one of January 10, 1901, has unusually ample macroseismic documentation. Macroseismic maps of June 11, 1895 earthquake were compiled by Leonhard and Voltz (1895, 1896), Dathe 1897 (see also Kozak 2015, this issue). The earthquake occurred north of the Marginal Sudetic Fault, in an area of the Ramzova Overthrust, separating the Sudetes into their Western and Eastern parts. This earthquake still remains the only known seismic event from this area.

Seismic activity of the Odra River Lineament (ORL) could be confirmed by 2 weak events recorded only in Wrocław, with intensity 4. The one on January 24, 1775 was reported by Professor Zeplichal (1775), a diligent and reliable witness of the event, with accurate time description of the main event and its foreshock and aftershock. These are the only known evidence of tectonic seismic events in the ORL but the induced seismicity in the Lubin cooper basin, located at the edge of the ORL, where the exploration started in 1968 is high. Gibowicz et al. (1979) supposed that a shock of $M_L = 4.5$ in the year 1979 was caused by mining activation of a pre-existing fault in the ORL zone.

3.4.4 Region IV: Outer Western Carpathians—Silesian Beskids, Cieszyn Silesia/Upper Silesia

The Outer Western Carpathian Arc, in the Silesian area where the Carpathians thrust over the Upper Silesian Block, was unusually active in the very short, 2-year period 1785–1786 when 3 earthquakes of $I_o = 6$–7 (August 22, 1785; February 27, 1786; December 3, 1786) were recorded, with exceptionally deep foci, reaching 35 km. Foci of the main earthquakes in the years 1785-1786 occurred in the area of the Upper Silesian Block, close to the tectonic juction of the Carpathian and Varicsan orogenic fronts (Guterch et al. 2015a, b, this issue). All other known earthquakes in the Western Carpathians are much shallower; their depths are up to 20 km, but mostly not exceeding 10 km (Kárník 1961; Procházková and Dudek 1980; Prochazkova 1990). There are no reports of weaker earthquakes which would correspond in number to the occurrence of the XVIII century events. Nowadays, the activity of the region is confirmed by two events recorded in May and November 1994, of $M_w = 2.4$ and 2.6 (Draber et al. 2000a, b; Geofyzikální ústav[3]).

3.4.5 Region V: Outer Western Carpathians—Beskid Sądecki and Beskid Niski (NE of the PKB)

Beskid Sądecki Mts. The historical activity is confirmed by just one event, of May 6, 1909, of weak credibility. In the years 1992–1993, two series of seismic events have been recorded in Beskid Sądecki, about 50 km east of the Peripieninian Lineament. Main earthquakes had intensities $I_o = 5$ on June 26,

[3] http://www.czechgeo.cz/gfu-bulletin/?do=process.

1992, and $I_0 = 7$ on March 1, 1993 (Guterch et al. 2000). Likewise the events along the PKP, the foci were of shallow depths, less than 5 km. It is supposed that the earthquakes originated in the flysh complex or in the border area of the metamorphic basement.

During the seismic monitoring carried out in the region since August 2008 until December 2012, four events were recorded, of magnitude $0.9 \leq M_L \leq 2.0$ (Trojanowski et al. 2015; Plesiewicz and Wiszniowski 2015, this issue). On April 1, 2014, an earthquake of $M_L = 2.8$, followed by 3 aftershocks during next 2 days, was recorded near Szczawnica just at the border area of the PKP and Beskid Sądecki Mts (Plesiewicz and Wiszniowski 2015, this issue).

Beskid Niski Mts. On September 25, 2007, a series of 12 seismic events were recorded near Grybów in Beskid Niski, the main event being of $M_W = 2.6$. Identification of these events was only possible thanks to the setting in 2004 of a local seismic station Stebnicka Huta (STHS) by the Institute of Geophysics, Slovak Academy of Sciences. No other seismic events have been known in the area (Wiejacz et al. 2008).

3.4.5.1 Carpathian Foredeep

Events felt in Kraków, the city mentioned over the centuries in the description of historical earthquakes, are most probably related to the perception of Carpathian earthquakes. There is no other confirmation on seismic activity in this area.

3.4.6 Region VI: Western Carpathians—Pieniny Klippen Belt (Geographically Named Podhale)

The most seismically active and best recognized in Poland are the Western Carpathians, along the Pieniny Klippen Belt (PKB), especially in the Orawa–Nowy Targ (ONT) Basin—intramontane depression, filled with Neogene and Quaternary sediments, at the northernmost part of the PKP (Guterch et al. 2005).

Foci of earthquakes recorded along the PKP are shallow, less than 5 km deep; macroseismic effects are strongly attenuated with distance and this is probably the reason why the events of $I_0 < 7$ have not been noted in the past. Macroseismic data of two seismic events recognized in the XVIII century are poor and their epicentral parameters are of small accuracy ($\Delta\varphi$, $\Delta\lambda > 50$ km, $\Delta I_0 > 1$). The first sufficiently well documented seismic event was recorded on March 25, 1935, of $I_0 = 7$. An earthquake of $M_W = 4.5$, $I_0 = 7$ was recorded in the ONT Basin on November 30, 2004, followed by a series of about 300 aftershocks until December 2005 (Guterch 2006; Wiejacz and Dębski 2009). Most houses in the epicentral area sustained damage of Grade 1 and many of Grade 2 in the EMS scale and a few buildings sustained damage of Grade 2–3 (Guterch 2006). Macroseismic maps of the main earthquake and 8 aftershocks, being a result of macroseismic survey done on-site in the epicentral area, are presented in Fig. 3.2 and 3.3.

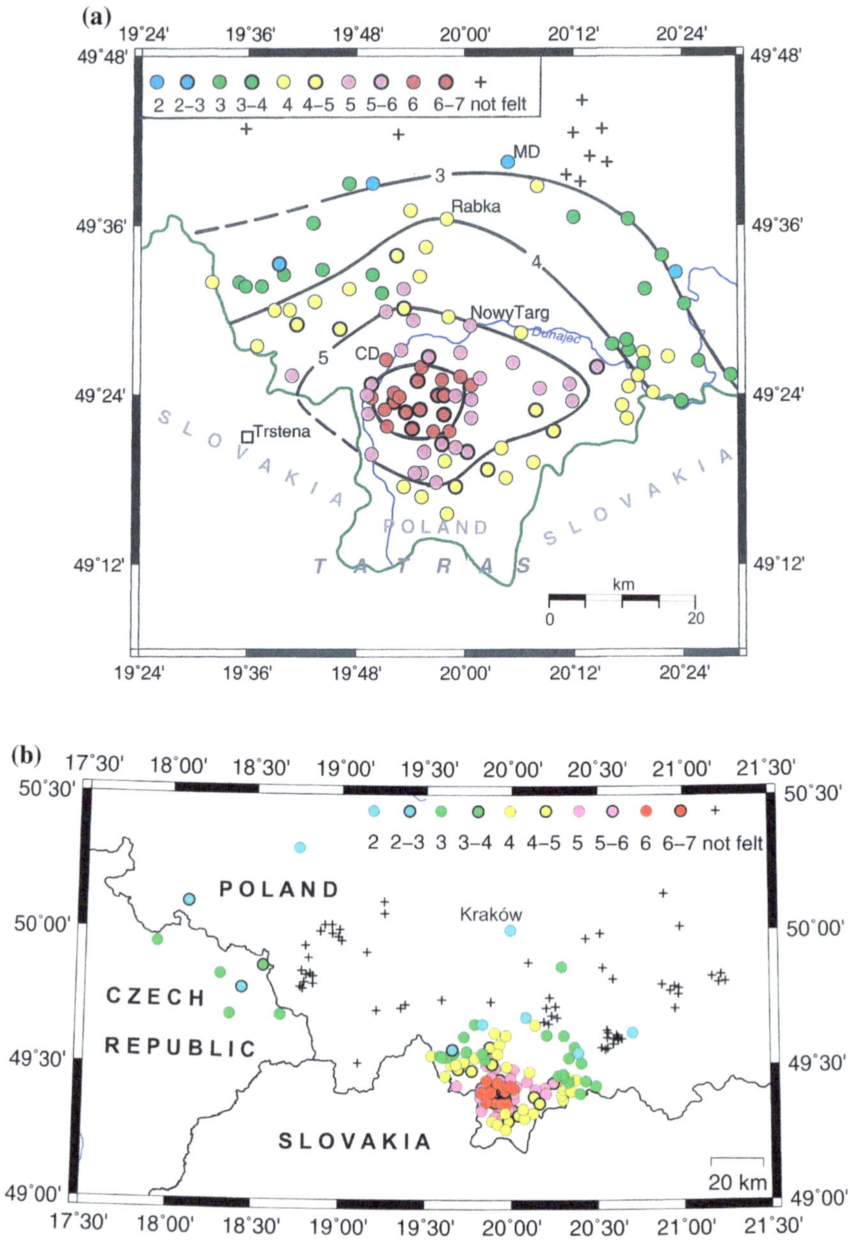

Fig. 3.2 Macroseismic maps in the EMS scale of the main earthquake of November 30, 2004, in the Orawa-Nowy Targ Basin, Pieniny Klippen Belt. **a** Macroseismic map of an area up to about 40 km from the epicenter (*MD* Mszana Dolna). **b** Intensity data points of the whole area where the earthquake was felt

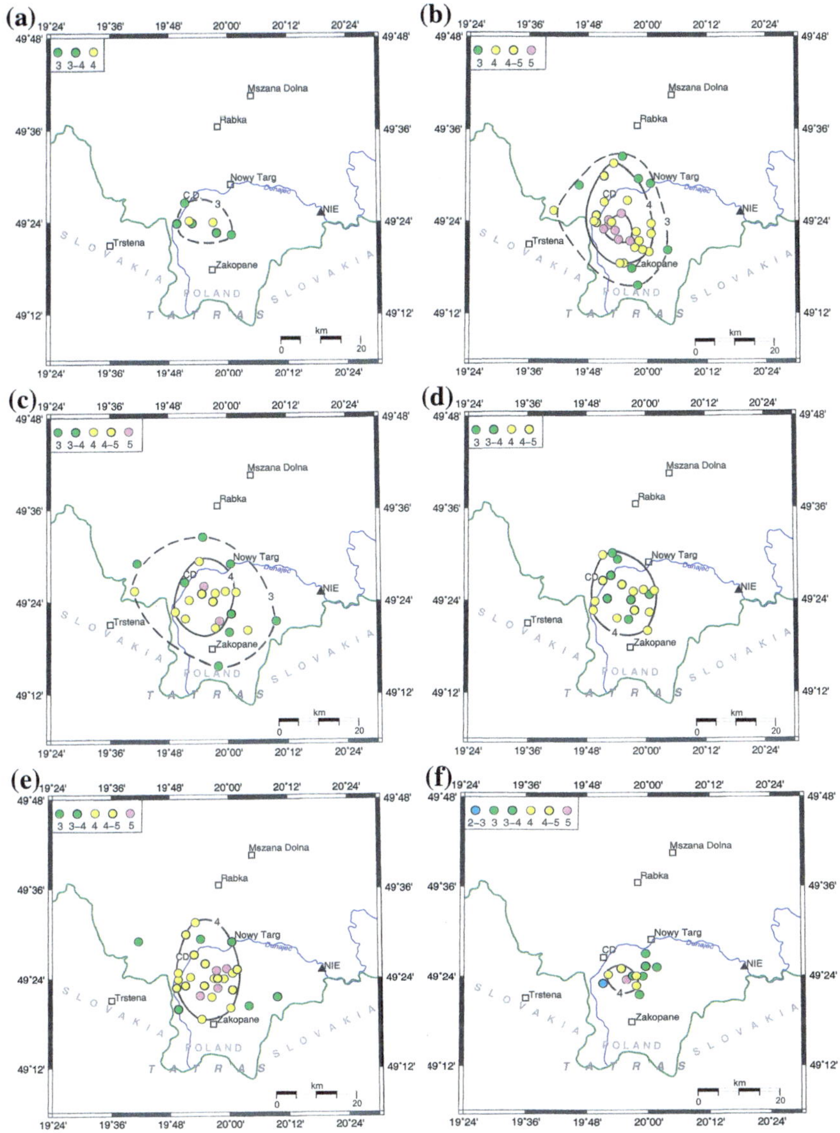

Fig. 3.3 Macroseismic maps in the EMS scale of the aftershocks of the main November 30, 2004, earthquake in the Orawa Nowy Targ Basin (*CD* Czarny Dunajec); **a** December 1, 2004, H = $23^h25^m15.6^s$, $M_W = 2.8$; **b** December 2, 2004, H = $18^h25^m39.2^s$, $M_W = 3.6$; **c** December 9, 2004, H = $01^h09^m05.8^s$, $M_W = 3.3$. **d** January 23, 2005, H = $23^h33^m20.0^s$, $M_W = 3.1$; **e** January 29, 2005, H = $17^h16^m56.2^s$, $M_W = 3.4$, **f** June 02, 2005, H = $07^h43^m27.2^s$, $M_W = 3.2$

Pomianowski (2003), by combining analyses of gravity and geoelectrical data, found that the ONT Basin is cut into blocks by two major fault systems: normal faults parallel or sub parallel to the PKP; and strike-slip faults mainly of NNW-SSE direction cutting and displacing the normal faults.

The normal fault plane mechanism of the main November 30, 2004, earthquake, made by the ETHZ Swiss Moment Tensor Solution[4] implies NNW oriented maximum extension. The strike and dip angles of faults in the Orawa-Nowy Targ Basin would suggest nodal plane of strike = 246°, rake = −80°, dip = 66° as the fault plane (Guterch 2006).

Tectonic activity since the Middle Miocene to Quaternary at the southern margin of the Orawa Depression, just in the epicentral area of the November 30, 2004, earthquake was revealed by studies of fractured clast (Tokarski and Zuchiewicz 1998; Zuchiewicz et al. 2002).

Seismic monitoring carried out by the Institute of Geophysics, Polish Academy of Sciences, along the PKP in Podhale revealed constant seismic activity: over 100 seismic events of magnitudes $0.5 \leq M_L \leq 2.3$ have been recorded over a period since August 2008 until December 2012 (Trojanowski et al. 2015). Completeness of the seismic catalogues and probability of magnitude exceedance was studied by Plesiewicz and Wiszniowski (2015, this issue).

Since the year 2011, weak seismic activity has been recorded at the south-western edges of the ONT Basin in the area of Czorsztyn Reservoir, on Dunajec river, associated with the water power station Niedzica. The reservoir was completed in 1996 and started to be filled up. A series of seismic events of a swarm type occurred in 2013; the strongest events were recorded on March 1, of $M_W = 3.4$ and March 2, of $M_W = 3.1$. (after Seismological Bulletins of the Institute of Geophysics PAS stations[5]). The events of $M_W \geq 2.6$ were felt by people, mostly around the reservoir, northwest of the dam, the strongest on March 1 at Kluszkowce 4° and Frydman 4°. Macroseismic data allow to assess intensity of the strongest events: March 1, 21:23:12.9 GMT, $I_o = 4$ EMS; March 2, 10:50:35.8 GMT, $I_o = 3$ EMS (see Table 3.1). It seems possible that the Czorsztyn Reservoir has triggered local tectonic activity in the ONT Basin.

3.5 Remarks and Conclusion

Seismic recognition of the area of Poland is extremely poor. There are areas known historically from the occurrence of earthquakes of $I_o = 6$–7 but these are single known events, never ever repeated there before or afterwards. It concerns mainly the Fore Sudetic Block (Paleozoic Platform), the wide Teisseyre-Tornquist Zone and the Precambrian Platform.

---

4 http://www.seismo.ethz.ch/moment_tensor/2004/041130_1718.ap.30_50s4k.

5 http://private.igf.edu.pl/~nlkozlak/.

Kaliningrad earthquakes in the year 2004 of M = 5.2 proved how little we know about seismicity of the Precambrian Platform. A series of aftershocks of Kaliningrad earthquakes, felt by people on the Sambia Peninsula, was not recorded instrumentally. Seismological stations are not able to record events of $M \ll 3$ at regional distances. The nearest seismic station Suwałki (SUW) set in the year 1995 was at a distance of 220 km from the epicenter.

Special attention is needed in seismic recognition of the Outer Western Carpathians in the Silesian area (region IV), the region of tectonic junction of the Variscan and Carpathian orogenic fronts. It is region where, in the 2-year period 1785–1786, three earthquakes of $I_o = 6–7$ occurred, two of them of $M_w = 5.3$ and h = 35 km (Guterch et al. 2015a, this issue).

The poor seismic recognition of Poland was the reason for putting forward, upon the initiative of Prof. Aleksander Guterch, a project of seismic monitoring of the country. The project entitled "Monitoring of Seismic Hazard in Poland" on the order of the Ministry of the Environment, was financed by the National Fund for Enviromental Protection and Water Management (contract no. 445/2007/Wn-07/F6-bp-tx/D). In the framework of the project, 24 special mobile seismic stations were constructed in the Institute of Geophysics, Polish Academy of Sciences. The project was in operation since July 2008 until December 2012 (Trojanowski et al. 2009). It has revealed a weak seismic activity in Beskid Sądecki (region V) and especially along the Pieniny Klippen Belt (region VI), where over 100 seismic events of magnitudes $0.5 \leq M_L \leq 2.3$ were recorded (Plesiewicz and Wiszniowski 2015, this issue). The occurrence of an earthquake of $I_o = 5$, $M_w = 3.5$ on January 6, 2012, in the Teisseyre–Tornquist zone near Jarocin, in an area where a seismic event of $I_o = 4–5$ occurred in the year 1824, confirmed the conclusion that seismic monitoring should be carried on, at least in areas where seismic events have been recorded in the past.

Acknowledgments I am grateful to Professor Jan Kozák for providing materials on historical earthquakes in Poland from his own collections and archives of the Geophysical Institute in Prague. I am grateful to my colleagues from the Seismological Department of the Institute of Geophysics for making available to me the data base on location and magnitudes of local earthquakes in Poland in the year 2013. I acknowledge the help of Edward Gaczyński with preparing the figures and tables for publication. Special thanks to Anna Dziembowska for encouragement, assistance and help in preparing the whole monograph.

References

AST (1778) Auszug Eines Schreibens AUS Tiefhartmannsdorf (1778) Ökonomische Nachrichten der Patriotischen Gesellschaft in Schlesien, 1778. In Commission bei Wilhelm Gottlieb Korn, sechster Band, Breslau, p 180

Bülow K (1928) Erdbeben Beobachtungen in Pommern, Abhandlungen und Berichte der Pommerschen Naturforschenden Gesellschaft. Stettin 9:159–172

Credner H (1901) Das Sachsische Schuttergebiet des Sudetischen Erdbebens vom 10. Januar 1901, Ber. d. math.-phys. Cl. d. k.k . Gesell. d. Wiss. zu Leipzig; Sitz. 4. Marz 1901, pp 83–103

Dathe E (1897) Das schlesisch-sudetische Erdbeben vom 11 Juni 1895, Abhandlungen d. k. preus. Geol. Landesanstalt, Neue Folge, Berlin, Heft 22

Długosz J (1873, 1877) Joannis Dlugossii seu Longini Canonici Cracoviensis Historiae Polonicae Libri XII. In: Przezdziecki A (ed) Joannis Długosz Senioris Canonici Cracoviensis Opera Omnia, Kraków, Cracoviae, ex Typographiae Kirchmayerianea, X, XI (1873), XIII (1877) (in Latin)

Dębski W, Guterch B, Lewandowska-Marciniak H, Labák P (1997) Earthquake sequences in the Krynica region, Western Carpathians, 1992–1993. Acta Geophys Pol 40:255–290

Draber D, Guterch B, Lewandowska-Marciniak H (eds) (1998) Seismological bulletin 1989–1990. Local earthquakes recorded by polish seismological stations, Publs Inst Geophys Pol Acad Sci B-17(283):129 p

Draber D, Guterch B, Lewandowska-Marciniak H (eds) (2000a) Seismological bulletin 1991–1992. Local earthquakes recorded by Polish seismological stations. Publs Inst Geophys Pol Acad Sci B-21(316):129 p

Draber D, Guterch B, Lewandowska-Marciniak H (eds) (2000b) Seismological bulletin 1993–1994. Local earthquakes recorded by Polish seismological stations. Publs Inst Geophys Pol Acad Sci B-22(331)

Draber D, Guterch B, Lewandowska-Marciniak H, Niewiadomski J (eds) (2003) Seismological bulletin 2001. Local earthquakes recorded by Polish seismological stations. Publs Inst Geophys Pol Acad Sci B-33(366):143 p

Fleszarowa RD (1926) Trzęsienie ziemi w Piotrkowskiem, Ziemia, no 11, pp 15–16

Garetsky R, Boborykin A, Yemeljanov A (1989) Historical earthquakes on the territory of Byelorussia and the Baltic republics. In: Sofia A (ed) Proceedings of European seismological commission, 23–28 August, 1988, Instituto Geográfico Nacional, Madrid, 217–230

Giardini D, Grünthal G, Shedlock KM, Zhang P (1999) The GSHAP global seismic hazard map. Annali di Geofisica 42(6) 1225–1228

Gibowicz SJ, Bober A, Cichowicz A, Droste Z, Dychtowicz Z, Hordejuk J, Kazimierczak M, Kijko A (1979) Source study of the Lubin, Poland, tremor of 24 Mar 1977. Acta Geoph Pol 27:3–38

Gregersen S, Wiejacz P, Dębski W, Domański B, Assinovskaya B, Guterch B, Mantyniemi P, Nikulin VG, Pacesa A, Puura V, Aronov AG, Aronova TI, Grunthal G, Husebye S, Sliaupa S (2007) The exceptional earthquake in Kaliningrad district, Russia, on 21 Sept 2004. Phys Earth Planet Int 164(1–2):63–74

Gränzer J (1901) Das sudetische Erdbeben vom 10. Jänner 1901, Mitteilungen des Vereines der Naturfreunde in Reichenberg, Jahrg. 32, Liberec (Reichenberg)

Grünhagen C (1870) Urkunden der Stadt Brieg, Urkundliche und Chronikalische Nachrichten über die Stadt Brieg, die dortigen Kloster, die Stadt und Stiftsgüter bis zur Jahre 1550. In: Codex Diplomaticus Silesiae, 9, Breslau, Biblioteka Uniwersytetu we Wrocławiu (Library of the Wrocław University) Syg. 15736 III

Grünthal G (1988) Erdbebenkatalog des Territorium der Deutschen Demokratischen Republik und abgrenzenden Gebiete von 823 bis 1984, Veröffentlichungen des Zentralinstituts für Phys der Erde, no 99

Grünthal G (1999) Seismic hazard assessment for Central, North and Northwest Europe: GSHAP region 3. Annali di Geofisica 42(6):999–1011

Grünthal G, Walström R, Stromeyer D (2009) The unified catalogue of earthquakes in central, northern and northwestern Europe (CENEC)–updated and expanded to the last millennium. J Seismol 13:517–541

Guterch B, Lewandowska-Marciniak H, Labák P, Niewiadomski J (2000) Macroseismic data of seismic events in 1992–1993, Krynica region, Western Carpathians. Publs Inst Geophys Pol Acad Sci B-21(316):133–147

Guterch B, Lewandowski-Marciniak H (2002) Seismicity and seismic hazard in Poland. Folia Quater 73:85–99

Guterch B, Lewandowska-Marciniak H, Niewiadomski J (2005) Earthquakes recorded in Poland along the Pieniny Klippen Belt, Western Carpathians. Acta Geophys Pol 53:27–45

Guterch B (2006) Seismic events in the Orawa-Nowy Targ Basin, Western Carpathians, November 30, 2004–December 2005, Acta GeodynGeomater 3(3) (143):85–95

Guterch B (ed) (2007) Seismological bulletin 2004. Local earthquakes recorded by Polish seismological stations. Publs Inst Geophys Pol Acad Sci B-40 (397)

Guterch B (ed) (2008) Seismological bulletin 2005. Local earthquakes recorded by Polish seismological stations. Publs Inst Geophys Pol Acad Sci B-41(399)

Guterch B (2009a) Macroseismic data of the Kaliningrad earthquakes recorded in Poland on September 21, 2004. In: AV Nikolaev (ed) Kaliningrad earthquake of 21st September, 2004, VSEGEI, Saint Petersburg, 97–99 (in Russian)

Guterch B (ed) (2009b) Seismological bulletin 2006. Local earthquakes recorded by Polish seismological stations. Publs Inst Geophys Pol Acad Sci B-42(410)

Guterch B, Lewandowska-Marciniak H, Mazur M (2015a) The earthquake in the Outer Western Carpathians, Poland, on December 3, 1786. In: Guterch B, Kozak J (eds) Studies of historical earthquakes in Southern Poland. GeoPlanet: earth and planetary sciences. Springer, Berlin, this issue

Guterch B, Lewandowska-Marciniak H, Mazur M (2015b) Source materials to the Outer Western Carpathian earthquake of December 3, 1786. In: Guterch B, Kozak J (eds) Studies of historical earthquakes in Southern Poland. GeoPlanet: earth and planetary sciences. Springer, Berlin, this issue

Hoff KEA (1824) Die Geschichte der durch Überlieferung nachgewiesenen natürlichen Veränderungen der Erdoberfläche, II Theil, Gotha, p 303

Hoff KEA (1841) Chronik der Erdbeben und Vulkanausbrüche, Gotha, 78, p 88 (in German)

Janczewski EW (1933) Ruchy sejsmiczne zauważone w Polsce w lutym 1932 roku. Posiedzenia Naukowe Państwowego Instytutu Geologicznego. Warszawa 33:70–72

Jeitteles HL (1860) Versuch einer Geschichte der Erdbeben in den Karpathen und Sudeten-Ländern bis zu Ende des achtzehnten Jahrhunderts, Zeitschrift d Deutschen Geol Ges, 12 Bd. 287–349

Kárník V, Michal E, Molnar A (1958) Erdbebenkatalog der Tschechoslowakei bis zum Jahre 1956, Geofysikalni Sbornik 1957. Travaux de l'Institut Geophysique de l'Academie Tchecoslovaque des Sciences 69:411–631

Kárník V (1961) Die Seismizität der Westkarpathen, Geofysikalni Sbornik 1960. Travaux de l'Institute Geophysique de l'Academie Tchecoslovaque des Sciences 134:265–283

Koláček F (1916) Zemětřesení na Moravě a v rakouském Slezsku, 23, Sborník České Společnosti, Praha, pp. 153–160, (in Czech)

Koláček F (1921) Seznam otřesu pozorowaných na pudě Republiky Československé, in: Hostinský B (ed) Publications de la Faculté des Sciences de l'Université Mssaryk 7, 10 p, Brno (in Czech)

Kozák J, Plešinger A (2003) Beginnings of regular seismic service and research in the Austro-Hungarian Monarchy: Part I. Stud Geophys Geod 47:99–119

Kozák J (2015) First macroseismic maps in southern Poland in late 19th century. In: Guterch B, Kozak J (eds) Studies of historical earthquakes in Southern Poland. GeoPlanet: earth and planetary sciences. Springer, Berlin, this issue

Kromer M (1611) Kronika polska biskupa warmińskiego ksiąg XXX, Kraków, reedited: 1857, p 360, 467

Laska W (1902) Die Erdbeben Polens, Mittheilungen der Erdbeben Commission, vol 8. Keis. Ak. d. Wiss. in Wien, pp1–36

Lehman FWP (1913) Erd und Seebeben in Pomern. Monatsblätter d. Ges., f. Pommersche Geschichte und Altertumskunde. 2:18–23

Leonhard R, Voltz W (1895) Das mittelschlesischen Erdbeben vom 11 Juni 1895. Jahresbericht d. Schles. Gesell. f. vaterl. Kultur, Naturw. Sektion II., Abt. XXIV, Breslau

Leonhard R, Voltz W (1896) Das mittelschlesischen Erdbeben vom 11 Juni 1895. Zeitschrift der Ges. f. Erdkunde zu Berlin, Bd. XXXI, 1–21

Lizurek G, Plesiewicz B, Wiejacz P, Wiszniowski J, Trojanowski J (2013) Seismic event near Jarocin (Poland). Acta Geoph 61(1):26–36

Málek J, Brož M, Stejskal V, Štrunc J (2008) Local seismicity at the Hronov-Poříčí fault (Eastern Bohemia). Acta Geodyn Geomater 5(2) (150), 171–175
Mazur M (1968) Wstępne informacje o serii wstrząsów tektonicznych w rejonie Zakopanego w marcu 1966 roku. Acta Geophys Pol 16:201–204
Miechowita M (1986) Chronica Polonorum Mathiae de Mechovia, Liber Tercius. In: Wietor H (ed) (1521) facsimile, Sł. Tabkowski (ed) (1986) Krajowa Agencja Wydawnicza, Kraków, p 157, 159 (in Latin)
Nikonov AA (2005) Were there worthy of note earthquakes and tsunamis within the South-Eastern Baltic area? Kaliningrad earthquake September 21, 2004, Workshop Materials, Tartu 23–25
Olczak T (1962) Sejsmiczność Polski w okresie 1901–1950. Acta Geophys Pol 10(1):3–11 (in Polish)
Olczak T (1966a) Sur la region epicentrale et la profondeur d'un foyer seismique du 3 XII 1786 dans les Carpathes occidentales polonaises. Acta Geophys Pol 14(1):61–67 (in French)
Olczak T (1966b) O Karpackim trzęsieniu ziemi z dnia 27 II 1786 roku i towarzyszących mu ruchach sejsmicznych na przedpolu Sudetów Wschodnich. Przegląd Geologiczny XI (XIX) (1) 3–21 (in Polish)
Olczak T (1965) Trzęsienie ziemi z dnia 22 VIII roku w Beskidzie Śląskim. Przegląd Geologiczny X(XVIII) (3–4) 227–237 (in Polish)
Pagaczewski J (1972) Catalogue of earthquakes in Poland in 1000–1970 years. Publs Inst Geophys Pol Acad Sci 51:3–36
Plesiewicz B, Wiszniowski J (2015) Seismicity of Polish part of the Western Carpathians in the light of recent data, In: Guterch B, Kozak J (eds) Studies of historical earthquakes in Southern Poland. GeoPlanet: earth and planetary sciences. Springer, Berlin, this issue
Pomianowski P (2003) Tectonics of the Orawa-Nowy Targ Basin—Results of the combined analysis of the gravity and geoelectrical data (in Polish, abstract and description of figures in English). Prz Geol 51:498–506
Pohludka S (1897) Erdbeben in Mahren und Schlesien, Mitteilungen des Naturwiss. Vereines in Tropau, no 5, III. Vereinsjahr, p 95 (in German)
Prochazková D (1990) Seismicity of Central Europe. Publs Inst Geophys Pol Acad Sci B-14(231):96 p
Procházková D and Dudek A (1980) Some parameters of earthquake originated in central and eastern Europe, Travaux Geophysiques XXVIII, 538:43–82
Remeš M (1925) Zemětreseni pozorovaná na Moravě a ve Slezsku, Věstnik Klubu přirodověd. v Prostějově za rok 1922–1925, Jg. XIX, Prostějov (in Czech)
Rocznik Magistratu Wrocławskiego (1878) In: Bielowski A (ed) Annales Wratislaviensis, Monumenta Poloniae Historica, Lwów, vol III, p 686 (in Latin)
Schenk V, Schenková Z, Kottnauer P, Guterch B, Labák P (2000) Earthquake hazard for the Czech Republic, Poland and Slovakia, Contribution to the ILC/IASPIE Global Hazard Assessment Program. Natural Hazards 21:331–345
Schenk V, Schenková Z, Kottnauer P, Guterch B, Labák P (2001) Earthquake hazards maps for the Czech Republic, Poland and Slovakia. Acta Geophys Pol 49(3):287–302
Schenková Z, Schenk V, Kottnauer P, Guterch B, Labák P (1999) Earthquake catalogue for the Czech Republic, Poland and Slovakia (manuscript)
Sieberg A (1940) Beitrage zum Erdbebenkatalog Deutschlands und angrenzenden Gebiete für die Jahre 58 bis 1799. Mitteilungen des Deutschen Reichs-Erdbebendienstes, Heft 2
Sponheuer W (1952) Erdbebenkatalog Deutschlands und der angrenzenden Gebiete für die Jahre 1800 bis 1899, Mitteilungen des Deutschen Erdbebendienstes, Heft 3, Deutsche Ak. d. Wiss. zu Berlin, 195 p
Sturm F (1903) Das Sudetische Erdbeben vom 10. Januar 1901. Neuen Jahrbuch f. Mineralogie, Geologie und Paläntologie, Beilage–Band XVI. pp 199–240, with 2 maps, E. Naegele, Breslau

Tokarski AK, Zuchiewicz W (1998) Fractured clasts in the Domański Wierch series. Contribution to structural evolution of the Orawa Basin (Carpathians, Poland) during Neogene through Quatternary times, Prz Geol 46(1):62–66

Tornquist A (1909) Über in Ostpreussen beobachtete Erdbebenerscheinungen an der Jahreswende 1908/1909. Schr Phys—Kon Ges 1:27–36 (in German)

Trojanowski J, Plesiewicz B, Guterch A, Grad M (2009) Monitoring zagrożenia sejsmicznego obszaru Polski, (Monitoring seismic hazards of Poland). Prz Geol 57:133–137 (in Polish)

Trojanowski J, Plesiewicz B, Wiszniowski J, Suchcicki J, Tokarz A (2012) Monitoring zagrożenia sejsmicznego obszaru Polski–okres realizacji 2008–2012. Raport końcowy (Final Report, unpublished)

Trojanowski J, Plesiewicz B, Wiszniowski J (2015) Seismic monitoring of Poland—description and results of temporary seismic project with mobile seismic network. Acta Geophs Pol (in press)

TZKP (1837-1858), manuscript, Trzęsienia ziemi w Królestwie Polskim 1837–1858(?), Acta Komisji Rządowej Spraw Wewnętrznych, Duchownych i Oświecenia Publicznego, inwentarz KRSW (Komisja Rządowa Akt Wewnętrznych), Archiwum Akt Dawnych w Warszawie (Central Archives of Historical Records in Warsaw) folia 7280, 33 p (in Polish)

Turno A (1824) Pamiętnik, Biblioteki Zakładu Narodowego im. Ossolińskich, Wrocław, księga VIII, p 40, syg 13814/III/8 (in Polish)

Wiejacz P (2006) The Kaliningrad earthquakes of September 21, 2004. Acta Geodyn Geomater 3(2) (142):7–16

Wiejacz P, Dębski W (2001) New observations of Gulf of Gdańsk seismic events. Phys Earth Planet Inter 123(2–4):233–245

Wiejacz P, Dębski W, Kotwicki A, Rudziński Ł (2008) Microearthquakes near Grybów, Poland, Western Carpathians. http://private.igf.edu.pl/~rudzin/pliki/ESC2008gry_post.pdf

Wiejacz P, Dębski W (2009) Podhale, Poland, earthquake of November 30, 2004. Acta Geophys 57(2):346–366

Woldřich (1901) Das Nord Ost–bohmische Erdbeben vom 10 Janner 1901. Sitzungber. d.k. Akademie d. Wiss. in Wien., math–naturwiss. Kl., Bd 110, Abt 1

Zedník J, Pospišil J, Ružek B, Horálek J, Boušková A, Jedlicka P, Skácelová Z, Nehybka V, Holub K, Rušajová J (2001) Earthquakes in Bohemia Massif and surrounding regions in 1995–1999. Stud Geophys Geod 45:276–282

Zerneke JH (1711) Historiae Thoruniensis Neufragae Tabulae, oder Kern der Thornischen Chronicke welches vom Jahre 1231 sten bis an das 1711te Jahr, Archiwum Państwowe (State Archives in Toruń), Toruń, Folia 3210

Zeplichal AM (1775) Geschichte einer in der Nacht des 24sten Jenners zu Breslau im Jesuitercollegio bemerkten Erderschütterung, Ökonomische Nachrichten der Patriotischen Gesellschaft in Schlesien, Dritter Band, Breslau, in Commission bei Wilhelm Gottlieb Korn

Zuchiewicz W, Tokarski AK, Jarosiński M, Mátron E (2002) Late Miocene to present day structural development of the Polish segment of the Outer Carpathians. EGU Stephan Mueller Special Publication Series 3:185–202

Zuchniak M, Niewiadomski J, Guterch B (eds) (2005) Seismological bulletin 2003. Local earthquakes recorded by Polish seismic stations. Publs Inst Geophys Pol Acad Sci B-37 (376)

Chapter 4
Seismicity of Polish Part of the Western Carpathians in the Light of Recent Data

Beata Plesiewicz and Jan Wiszniowski

Abstract This chapter presents our up-to-date knowledge on seismicity of the Polish part of the Western Carpathians. It also gives a list of seismic sites and description of measurement procedures applied in the discussed region thus far. We mainly focus on seismicity of the Podhale region, the Pieniny Klippen Belt and the Beskids. At the moment there are 5 seismic stations in the Podhale region, which record the permanent weak seismic activity at the level of $0.5 \leq M_L \leq 2.3$. In this paper we estimated the completeness of catalogs and the distribution of magnitude exceedance probability for events in the Podhale region.

Keywords Seismicity · Western Carpathians · Podhale · Seismic network · Catalog completeness · Probability distribution

4.1 Introduction

The Carpathians are part of the Alpine Mediterranean Belt. The seismicity of the Carpathians is mainly observed in the Vrancea zone in Romania. Three strong earthquakes, with magnitudes larger than 6.5, occurred there in the last 30 years, and earthquakes with magnitudes of 5 occur almost every year (Zsíros 2003). Another seismic area in the Carpathians is the Pannonian Basin. Magnitude 6 earthquakes occur once in about 100 years, while magnitude 5 events occur every 20 years on the average (Tóth et al. 2008).

The Western Carpathians region is the northwesternmost part of the Carpathian mountain chain. It is located mostly on the territory of Slovakia and Poland, and

B. Plesiewicz (✉) · J. Wiszniowski
Institute of Geophysics, Polish Academy of Sciences, Warsaw, Poland
e-mail: bples@igf.edu.pl

J. Wiszniowski
e-mail: jwisz@igf.edu.pl

B. Guterch and J. Kozák (eds.), *Studies of Historical Earthquakes in Southern Poland*, GeoPlanet: Earth and Planetary Sciences,
DOI 10.1007/978-3-319-15446-6_4

partly in the Czech Republic, Hungary and Austria. The Western Carpathians have a significant seismicity. Epicenters of earthquakes, that occurred manly on the territory of Slovakia, have been documented in the Labák and Brouček's (1996) catalog since the year 1034. The most seismically active in the Western Carpathians are the regions of Žilina, Dobrá Voda (the best spatially defined focal zone), Malé Karpaty Mts, Central Slovakia, Komárno and Slanské Vrchy Mts (Slovakia). The three largest earthquakes were those of 5 June 1443 in Central Slovakia, 28 June 1763 in Komárno and 9 January 1906 in Dobrá Voda, all with epicentral intensity $I_0 = 8–9$ (MKS) (Kováč et al. 2002). Earthquakes from Žilina have been known since the XVII century, where the strongest event was that of 16 November 1613 with $I_0 = 8$ (Labák and Brouček 1996), causing damage to buildings in Bratislava. Another felt earthquake in this region was that of 15 January 1858 with epicentral intensity of about 8 (EMS) (Zsíros 2005).

The Western Carpathians are divided into three subprovinces: Outer, Central and Inner. The territory of Poland covers in part the Outer and Central Western Carpathians, and also in part the Outer Eastern Carpathians. In the area of Poland, the Western Carpathians region is the most seismically active region of the country, as known since the XVIII century. The strongest events occurred on 11 March 1717, 23 March 1935 and 30 November 2004 in the Podhale area (Laska 1902; Pagaczewski 1972; Guterch 2009). Their epicetral intensity was 7.

The instrumental seismic measurements have been carried out in Poland since the beginning of the last century. The recent data cover permanent digital measurements by the Polish Seismological Network (PLSN) and measurements by mobile seismic monitoring stations. Since 2008, the Institute of Geophysics, Polish Academy of Sciences (IGF PAS) has begun seismic measurement campaign focused on the observation of local seismicity and neotectonic recognition of the Western Carpathians. The recording of permanent seismicity made it possible to verify the completeness of the catalogs and determine the seismic hazard parameters of the Podhale region in the Central Western Carpathians.

4.2 Seismic Measurements in Poland

On the territory of the present-day Poland, seismic measurements have been conducted since the beginning of XX century. The first seismological observatory in this area was established by Professor M.P. Rudzki at the Jagiellonian University in Cracow (KRA) in 1903. It was equipped with two horizontal seismographs of Bosch-Omori type. Its high-quality data over the period 1903–1916 are still used in the world literature. Another seismological observatory with long tradition was the Silesian Geophysical Observatory at Racibórz (RAC), which started working in 1928, established (in then German town Ratibor) and directed by Professor Karl Mainka, inventor of the MAINKA seismographs. Since 1953 the station RAC, managed by the IGF PAS, has been continuing its work. There was also the “Erdbebenwarte Breslau-Krietern” station in Wrocław (BRE) (then German city

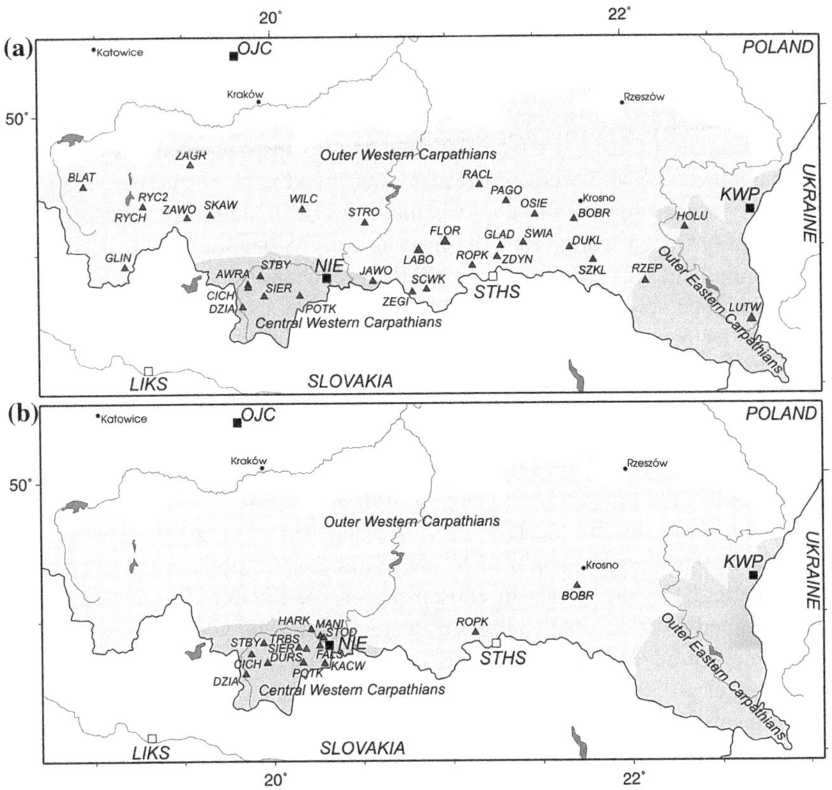

Fig. 4.1 Location of seismic stations in the Polish part of the Western Carpathians: **a** MSHTP project: 2008–2012, **b** PGMP-I project: from 2013. *Triangles*—temporary seismic network, *squares*—National Seismic Networks (*black*—Polish, *white*—Slovakian)

Breslau), established in 1909, operative (with some breaks) until the Second World War. We should also mention the Seismological Station in at the Polytechnic in Lwów (now Lviv in the Ukraine), established as early as 1901 (for more information about the history of seismology in Poland see, e.g., Maj and Teisseyre 2014).

Since 1953, the seismic measurements on the territory of Poland have been managed by the Institute of Geophysics PAS; until the 1990s these were analog recordings. Digital recording of seismic signal started in fact in 1988. The Polish Seismological Network (PLSN) consists of 7 broadband and 2 short-period stations. The seismic data are exchanged online with foreign institutions and seismological centers.

Local seismicity in the Western Carpathians in southern Poland is monitored mainly by seismic stations Niedzica (NIE), Ojców (OJC) and Kalwaria Pacławska (KWP) (Fig. 4.1). The NIE was established in the mid-1960s on the historical castle in Niedzica. The seismic noise was related to human activity in the castle. The station was moved in 1994 to a new location on a mountain top about 2 km away from the dam on the river Dunajec, and started operating as a short-period digital station. Despite the proximity of the power plant, the recording conditions at NIE were remarkably good; however, the triggered mode of recording limited the detection of small events. In September 2005 there started a continuous recording of 20 sps (samples per second) data, whereas 100 sps data were still recorded in triggered mode. In 2013 it started continuous 100 sps data recording. In April 2009, seismometers SM-3 were replaced by the broadband seismometers STS-2. Station OJC was put up in 1990 and it is the best Polish station in terms of noise conditions. It has been working all the time as a digital station. It has been using different equipment, but since August 1999 the broadband seismometer STS-2 has been applied. The seismic station KWP was put up in June 1999 in cooperation with GEOFON Network of the GeoForschungsZentrum Potsdam (Germany). It has been equipped with the broadband seismometers STS-2 and performed continuous data recording, initially at a rate of 20 sps and from May 2010 at a rate of 100 sps. In addition, the seismic monitoring is supported by Slovak seismic stations LIKS (Likavka) and STHS (Stebnická Huta).

In 2008–2012, the project *Monitoring of Seismic Hazard of Territory of Poland* (MSHTP) was launched by the IGP PAS (Trojanowski et al. 2012). The goal of the project was to complement PLSN with temporary mobile seismic network to record natural seismic events in chosen regions of Poland. The stations' sites were determined after detailed analysis of historical seismic activity. The MSHTP project administered 24 seismic stations over a period of 5 years in two stages (Trojanowski et al. 2015). In the first stage of the project, since mid-2008 to mid-2010, the measurements were made in southern Poland, with special attention to the Western Carpathians. The recording was made in 33 sites. In the second stage, since mid-2010 to the end of 2012, seismic monitoring was made in central and southern Poland, while the Western Carpathians region was only monitored by the five stations: CICH, DZIA, SIER, STBY, and POTK (Fig. 4.1a, b).

The seismic network in Podhale continued working after the end of the MSHTP project. These measurements have been run by the IGP PAS within the *Permanent Geodynamical Monitoring of Poland—stage I* (PGMP-I) project. The period of realization of the PGMP-I project is 2013–2015. Seismic monitoring consists of 20 mobile stations in Poland, 14 being located in the Western Carpathians. The location of stations in the Western Carpathians and duration of their operation in the framework of MSHTP and PGMP-I projects is shown in Fig. 4.1a, b and Appendices A and B.

4.3 Seismicity of the Western Carpathians in Poland

4.3.1 Outer Western Carpathians—Carpathians Foothills and Beskids

The oldest historical earthquake in the Outer Western Carpathians, with epicentral intensity $I_0 = 4$, was located in the Carpathians Foothills in 1857 (Guterch 2009).

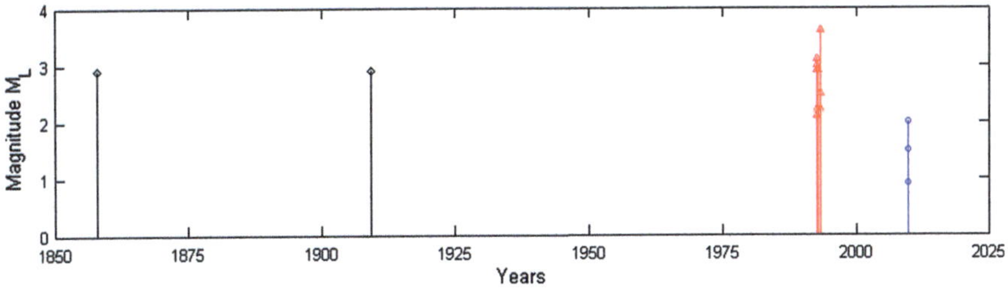

Fig. 4.2 Time sequence of earthquakes around Krynica. *Diamonds*—historical earthquakes, *triangles*—earthquakes recorded instrumentally by the PLSN, *circles*—earthquakes recorded by the MSHTP project

The Western Beskids is the region in which the historical seismic activity is poorly documented and can be confirmed only by one earthquake of 06.05.1909 with $I_0 = 4$. In the years 1992–1993, a series of shocks took place near Krynica (Figs. 4.2 and 4.4). The magnitudes of main earthquakes were $M_L = 3.6$ (29.06.1992) and $M_L = 4.0$ (01.03.1993) (Guterch 2009). For the stronger event, small construction damages were reported near the epicenter. Each of these events was accompanied by weaker foreshocks and aftershocks (Guterch et al. 2000; Guterch 2009). Seismic activity of the Krynica region was confirmed during the MSHTP project. Three earthquakes were registered. Two events occurred on 4 September 2009: the first with $M_L = 0.9$ and the other with $M_L = 1.5$. The strongest one, of 23 October 2009, had $M_L = 2.0$.

Seismological bulletins of Institute of Geophysics PAS exhibit in the period 2001–2002 [see Draber et al. 2003b, 2004) a series earthquakes of local magnitude about 2 on the border of the Western Beskids and the Pieniny Klippen Belt (Figs. 4.3 and 4.4). During the PGMP-I project, four events were registered in this region. The first and strongest event ($M_L = 2.8$) was on 01.04.2014 and was felt by some Szczawnica inhabitants. Later the same day, another event ($M_L = 1.8$) was registered, which was followed by two weaker earthquakes in next few days.

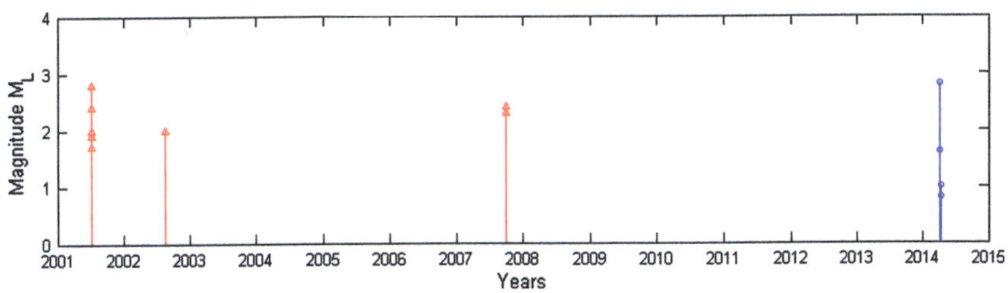

Fig. 4.3 Time sequence of earthquakes in the area between Szczawnica and Grybów. *Triangles*—earthquakes recorded instrumentally by the PLSN, *circles*—earthquakes recorded by the PGMP-I project

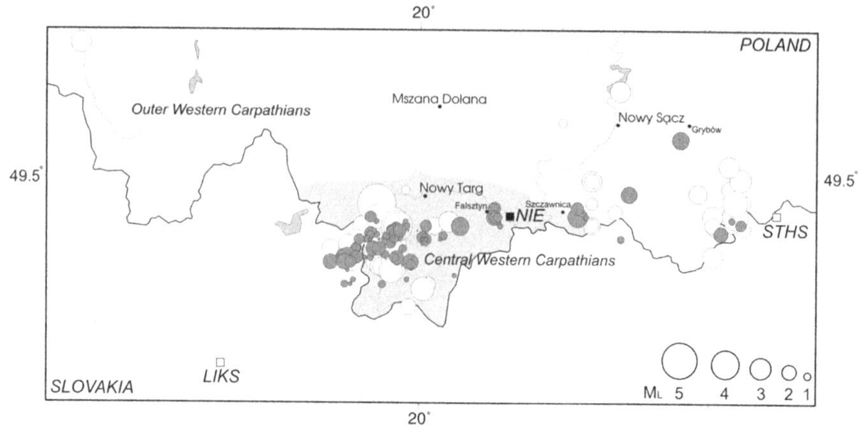

Fig. 4.4 Earthquakes in the Carpathians in Poland over the period from 1716 to 2007 (*white circles*) and recent events from 2008 to 2014 (*gray circles*)

The region of Central Beskids is very poorly explored. The reports on its seismicity are of weak reliability (Pagaczewski 1972). However, between 24 and 27 September 2007, seismic stations in Poland and Slovakia registered series of weak events in the neighborhood of Grybów town. On 24 and 25 September 2007 there were 13 events and on 27 September 2007 other two weak events which were recorded by STHS only. No felt reports have been known. Locations were only possible for the two most intense events from area of Grybów (Wiejacz et al. 2008): $M_W = 2.3$ at 16:27 on 25.09.2007, and $M_W = 2.6$ at 21:32 on 25.09.2007 (Fig. 4.4). There were no seismic events registrations from this area during the MSHTP and the PGMP-I projects. However, due to a widespread area of station sites, the detection threshold was at the level of $M_L = 1.5$.

4.3.2 *Central Western Carpathians—Podhale and Pieniny Klippen Belt*

Podhale is a relatively well examined seismic region in the Western Carpathians, where historical events of epicentral intensity I_0 in the range of 6–7 are known since the XVIII century (Table 4.1). The strongest instrumentally recorded shock in Podhale took place on 30 November 2004 ($M_L = 4.3$) in the Orava–Nowy Targ Basin (Guterch 2005). After the main shock, a series of aftershocks, which lasted for almost one year, took place. Because of the lack of seismic stations in this area, focal parameters could be calculated only for shocks of magnitudes greater than 2.0. Other historical earthquakes in the region are also known: $I_0 = 7$ on 23 March 1935 and on 11 March 1717, and weaker events that were instrumentally recorded since the late 1980s (Table 4.1).

Table 4.1 Historical events in the Podhale region (mainshocks, Guterch 2014)

Date	Longitude [°N]	Latitude[°E]	Epicentral intensity I_o	Magnitude M_L
1716	19.90	49.40	6	3.9
1717.03.11	19.90	49.40	7	4.4
1935.03.23	19.85	49.45	7	4.4
1942.03	19.90	49.40	4–5	3.0
1966.03.17	19.90	49.30	4–5	2.8

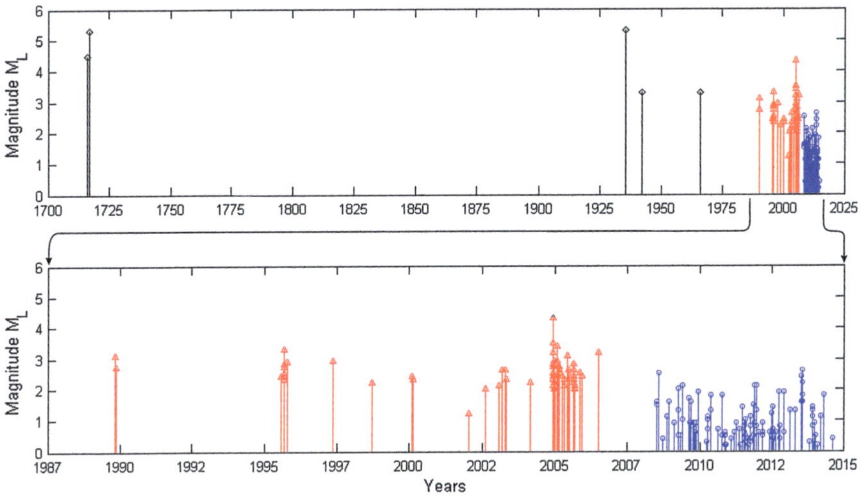

Fig. 4.5 Time sequence of all earthquakes in the Podhale region. *Diamonds*—historical earthquakes, *triangles*—earthquakes recorded instrumentally by the PLSN, *circles*—earthquakes recorded by the MSHTP and the PGMP-I projects

Podhale is the only region in Poland which shows constant natural seismic activity. During the MSHTP and the PMGP-I projects, between July 2008 and May 2014, about 150 earthquakes were registered in Podhale, 100 of them being located (Fig. 4.4). Moreover, in the area of Czorsztyńskie Lake, in November and December 2011, swarm events were registered in the Pieniny Klippen Belt area. Epicenters were localized in the Pieniny Klippen Belt near Falsztyn village, about 2 km from the Seismological Observatory NIE, where seismic activity has not been known so far.

The seismic monitoring network in Podhale registers, on the average, a natural seismic earthquake of $0.5 < M_L < 2.3$ once a month. These are weak events, at the level of noise. Figure 4.5 presents time sequence of all earthquakes in Podhale since 1700 until May 2014.

4.4 Distribution of Magnitudes in Podhale, Western Carpathians

Podhale is the only region in Poland in which a sufficient number of seismic events was recorded and the historical seismicity was documented in order to determine the probability of occurrence of shocks. To this end, it was necessary to use all the available information on the events in Podhale in various times. We can specify three different catalogs of seismic events in the Podhale region, including:

i. Historical events noticed up to the 1970s. The catalog includes only extreme events (Guterch 2014).
ii. Events recorded instrumentally by the PLSN from the late 1980s.
iii. Events recorded by seismic network in Podhale from the mid-2008 to the first three months of 2014.

The historical catalog includes 5 events noticed in the period 1716–1966 (Table 4.1).

The catalogs of instrumentally recorded events have different magnitude completeness thresholds. In order to determine the magnitude distribution, it was necessary to define the magnitude completeness threshold of these catalogs.

4.4.1 Completeness of Catalogs

Magnitude of completeness (M_C) of earthquake catalogs is the lowest magnitude at which all the earthquakes are detected in selected space and time volume. The completeness of catalogs can be estimated based on shape of Frequency-Magnitude Distribution (FMD) of detected events in the selected region as well as by other techniques (Mignan and Woessner 2012). The following methods were applied: Maximum Curvature (MAXC) method (Wyss et al. 1999; Wiemer and Wyss 2000), Entire Magnitude Range (EMR) method (Woessner and Wiemer 2005) and the *b*-value stability (MBS) method (Cao and Gao 2002).

The Maximum Curvature method estimates M_C as the point of the maximum curvature by computing the maximum value of the first derivative of the frequency-magnitude distribution. In practice, this matches the group of magnitudes with the highest frequency of events in the FMD. This technique requires fewer events than other techniques to reach a stable result, however it underestimates sometimes the M_C value (Wiemer and Wyss 2000; Woessner and Wiemer 2005).

The EMR technique estimates M_C using the entire magnitude range with events below the M_C. Woessner and Wiemer (2005) proposed a model consisting of two parts: the Gutenberg-Richter law for the complete part, and the cumulative normal distribution for the incomplete part of the FMD. They tested the lognormal and Weibull distributions as well.

Cao and Gao (2002) estimated Mc using the stability of the *b*-value of Gutenberg-Richter (Gutenberg and Richter 1944) model as a function of cut-off magnitude M_{Co}, named MBS by Woessner and Wiemer (2005). The M_C is defined as the magnitude for which the changes in b-value (Δb) are smaller than 0.03. It is based on the assumption that the estimated value of b increases for $M_{Co} < M_C$ and remains constant for $M_{Co} > M_C$. This method does not produce good results in case of high variability of the FMD. Woessner and Wiemer (2005) used the b-value uncertainty δb according to Shi and Bolt's (1982) criterion:

$$\delta b = 2.3b^2 \sqrt{\frac{\sum_{i=1}^{N} (M_i - \langle M \rangle)^2}{N(N-1)}},$$

where $\langle M \rangle$ is the mean magnitude and N is the number of events. In case of events in the Podhale region, the variability of the FMD is rather high, because of a small number of events.

In the years 1989–2007, seismic stations in Poland recorded 75 earthquakes (Draber et al. 1998, 2001, 2002, 2003a,b, 2004; Guterch 2006, 2007, 2009). After removing aftershocks of the event on November 30, 2004, there remain 50 events. They are shown in Fig. 4.6. To determine M_C the MAXC, EMR and MBS methods were applied. M_C values calculated by different methods and minimum recorded magnitude (M_{min}) are presented in Table 4.2. Basing on Table 4.2 and the FMD (Fig. 4.7a) we can assume that the catalog of events recorded instrumentally by the PLSN is complete from magnitude 2.4.

In the years 2008–2014, more than 100 seismic events with magnitude M_L from 0.2 to 2.4 was detected and localized in the Podhale region (Fig. 4.8). The best detection of small events is inside the network of seismic stations and outside it is worse. On the basis of values of M_C presented in Table 4.2, spatial distribution of magnitude of the recorded events (Fig. 4.8) and the shape of the FMD (Fig. 4.7b), it was assumed that the catalog of events is complete from magnitude 0.8.

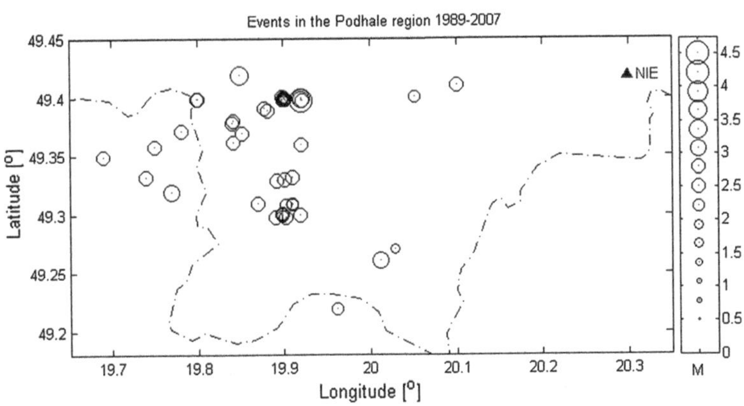

Fig. 4.6 Main seismic events in Podhale recorded in the years 1989–2007

Tab. 4.2 Magnitudes of completeness (M_C) of two catalogs of events (periods 1989–2007 and 2008–2014) estimated by various methods

Technique	M_C	
	1989–2007	2008–2014
MAXC	2.4	0.6
EMR	2.4	0.8
MBS	1.4	0.3
M_{min}	1.2	0.2

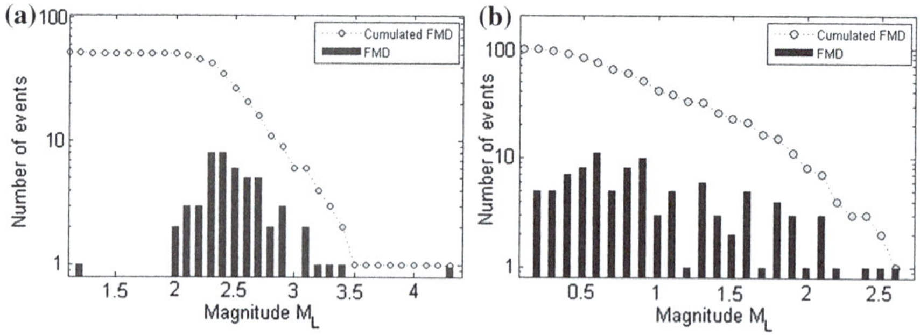

Fig. 4.7 The frequency-magnitude distribution (FMD) of events recoded in the Podhale region in the period 1989–2007 (**a**) and 2008–2014 (**b**)

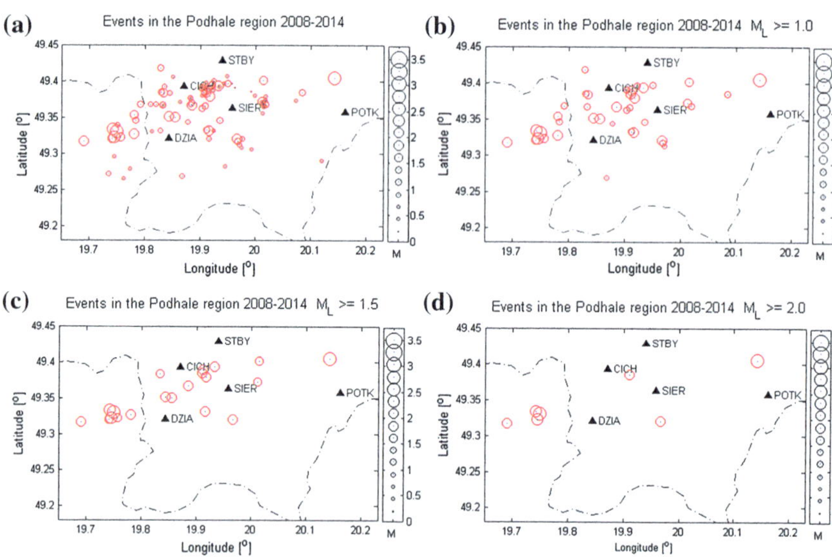

Fig. 4.8 Seismic events in Podhale in the period 2008–2014 for different cut-off magnitudes: **a** all events, **b** $M_L \geq 1.0$, **c** $M_L \geq 1.5$, and **d** $M_L \geq 2.0$

4.4.2 The Magnitude Exceedance Probability in the Podhale Region

The probability distribution of magnitude in the Podhale was calculated on the basis of estimation of earthquake hazard parameters from incomplete data catalogs

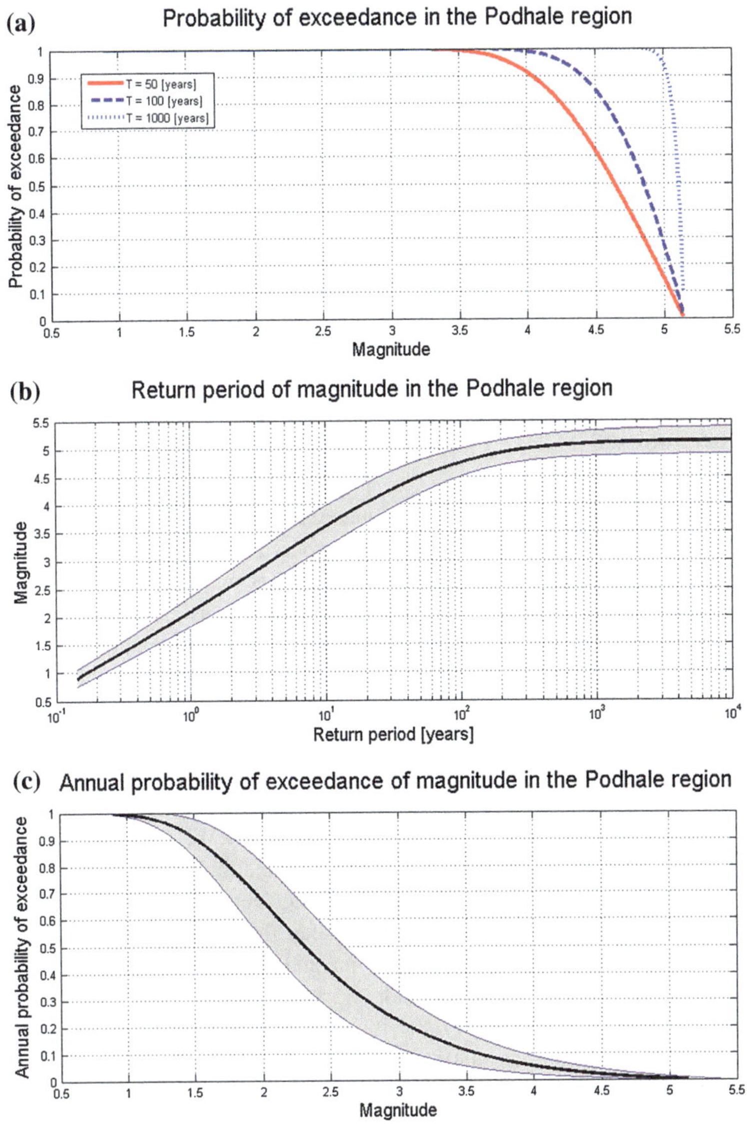

Fig. 4.9 The distributions of magnitudes in the Podhale region: **a** probability of exceeding the magnitude within the period of: 50 years—*solid line* (*red*), 100 years—*dashed line* (*purple*), 1,000 years—*dotted line* (*blue*); **b** return period versus magnitude; **c** annual probability of exceedance of magnitude; *gray* color in **b** and **c** indicates the range of uncertainty

(Kijko and Sellevoll 1989, 1992). We used three seismicity catalogs of Podhale: historical, catalog for 1989–2007, and catalog for 2008–2014. The maximum regional magnitude in the Podhale region was determined by Kijko-Selevoll-Bayes method (Kijko and Singh 2011). Calculations were made by HA2 program shared by the author, Prof. A. Kijko (Kijko 2010). The magnitude probability distribution parameters were:

$$\begin{aligned} &\boldsymbol{\beta} = 1.76 \pm 0.18 && (\boldsymbol{b} = 0.77 \pm 0.08) \\ &\boldsymbol{\lambda} = 9.795 \pm 2.457 && (\text{for } \boldsymbol{m}_{\boldsymbol{min}} = 0.90;\ \boldsymbol{a} = 1.67 \pm 0.11) \\ &\boldsymbol{m}_{\boldsymbol{max}} = 5.11 \pm 0.51 && (\text{for } \boldsymbol{m}_{\boldsymbol{max\,obs}} = 5.00 \pm 0.50) \end{aligned}$$

The distributions of magnitudes in the Podhale region is presented in Fig. 4.9. The probabilities of exceeding a magnitude within a defined period of time (50, 100, 500 years) are shown in Fig. 4.9a. The probability of exceeding the magnitude 5 is greater than 0.1 even for the shortest period (50 years). The return period of the particular magnitude is shown in Fig. 4.9b. Annual probability of exceeding the magnitude is presented in Fig. 4.9c. Probability of occurrence of events of $M_L = 1.5$ in one year is bigger than 0.9.

4.5 Conclusions

The Western Carpathians area had been monitored by the seismic station in Niedzica and additionally by stations of Slovak Seismological Network since 1960. However, the increase in detections of events in Podhale coincides with the expansion of PLSN, which began in the late 1980s, as well as the start of digital seismic measurements. This enabled the creation of complete catalog of seismic events with magnitude above 2.4 in this period. In the years 2008–2014, the mobile seismic monitoring of the Western Carpathians was carried out. Data from the seismic networks in the region significantly increased the amount of information on seismicity. The seismic network PLSN, as well as a temporary networks in the MSHTP and PGMP-I projects allowed for moving from macroseismic analysis in the region to registration of microevents.

The constant tectonic seismic activity in the Central Western Carpathians was detected in the area of Podhale. On the average, one shock a month was recorded in this region. Occasional events appeared in the Outer Western Carpathians as well. Basing on recent data we can conclude that they are grouped in two areas: around Krynica town and between Szczawnica and Grybów towns.

Owing to the monitoring of Podhale it was possible to create a complete catalog starting with magnitude of 0.8. This rendered us a possibility of attempting an evaluation of magnitude distribution. However, the duration of recording in the Podhale region was too short to get a sufficient number of strong events. Therefore, for estimating the magnitude distribution we had to use both the data from historic earthquake catalogs and those from catalogs of events recorded by PLSN. The magnitude distributions obtained are important from the engineering point of view, since it follows from them that the probability of the occurrence of an magnitude 5 event within 50 years is greater than 0.1, which should be accounted for, e.g., when buildings are designed.

However, to make a full assessment of seismic hazard of the Podhale region, a long-term continuation of measurements is indispensable, since no stronger events have been recorded in the period examined thus far. First of all, there is a lack of records of stronger ground motion amplitudes for estimating the Ground Motion Prediction Equations (GMPE). It is also advisable to further develop the network of seismic stations. This will enable determining the focal mechanism of the events in Podhale, and getting a better insight into the neotectonics of the area.

Acknowledgments The project *Monitoring of Seismic Hazard of Territory of Poland* (MSHTP) was carried by the Institute of Geophysics, Polish Academy of Sciences, in the period 2007–2012 on the order of the Ministry of the Environment and financed from funds of the National Fund for Environmental Protection and Water Management (contract No. 445/2007/Wn-07/F6-bp-tx/D).

The project *Permanent Geodynamical Monitoring of Poland—stage I* (PGMP-I) is carried by the Institute of Geophysics, Polish Academy of Sciences, in the period 2013—2015 on the order of the Polish Geological Institute—National Research Institute (contract No. E2-240-67-2013). This work was partially supported within statutory activities No 3841/E-41/S/2014 of the Ministry of Science and Higher Education of Poland.

Appendixes

Appendix A: List of seismic stations in the Western Carpathians—*Monitoring of Seismic Hazard on the Territory of Poland* project (in alphabetical order)

No.	Station code	Place	Registration period	Latitude [°N]	Longitude [°E]
1	AWRA	Ciche	2008.06.17–2008.08.10	49.40	19.87
2	BLAT	Błatnia	2009.05.06–2009.10.02	49.75	18.94
3	BOBR	Bóbrka	2008.09.09–2010.05.19	49.62	21.71
4	CICH	Ciche	2008.08.11–2012.12.31	49.39	19.87
5	DUKL	Dukla	2008.06.19–2008.08.06	49.52	21.68
6	DZIA	Dzianisz	2010.10.27–2012.12.31	49.32	19.84
7	FLOR	Florynka	2008.06.18–2008.09.04	49.55	20.98
8	GLAD	Gładyszów	2009.10.26–2010.01.09	49.53	21.29
9	GLIN	Glinka	2009.09.17–2010.04.14	49.46	19.18
10	HOLU	Hołuczków	2008.07.17–2010.06.08	49.58	22.33
11	JAWO	Jaworki-Biała Woda	2008.06.18–2008.07.20	49.41	20.57
12	LABO	Łabowa	2008.09.10–2010.05.19	49.52	20.83
13	LUTW	Lutowiska	2008.07.18–2010.05.31	49.24	22.69
14	OSIE	Osiek Jasielski	2008.06.19–2008.09.03	49.64	21.49
15	PAGO	Pagorzyna	2008.09.09–2010.05.27	49.69	21.33
16	POTK	Potok	2010.10.26–2012.12.31	49.36	20.16
17	RACL	Racławice	2008.08.07–2008.08.30	49.75	21.18

(continued)

(continued)

18	ROPK	Ropki	2008.08.09–2010.06.01	49.46	21.13
19	RYCH	Rychwałdek	2008.11.01–2009.05.05	49.68	19.28
20	RYC2	Rychwałdek	2009.05.05–2010.05.17	49.68	19.28
21	RZEP	Rzepedź	2008.07.19–2010.05.31	49.39	22.10
22	SCWK	Szczawnik	2008.08.09–2008.10.29	49.38	20.87
23	SIER	Sierockie	2008.08.11–2012.12.31	49.36	19.96
24	SKAW	Skawica	2008.07.20–2010.05.12	49.65	19.66
25	STBY	Stare Bystre	2008.07.20–2012.12.31	49.43	19.94
26	STRO	Stronie	2008.09.13–2010.06.01	49.62	20.53
27	SWIA	Światkowa Wielka	2008.08.08–2008.09.04	49.54	21.42
28	SZKL	Szklary	2008.08.07–2010.05.31	49.47	21.81
29	WILC	Wilczyce	2008.08.11–2010.05.31	49.67	20.18
30	ZAGR	Zagórze	2008.07.21–2010.06.16	49.83	19.55
31	ZAWO	Zawoja	2008.06.17–2008.07.20	49.64	19.53
32	ZDYN	Zdynia	2009.09.30–2009.10.26	49.49	21.27
33	ZEGI	Żegiestów	2008.10.29–2010.05.31	49.37	20.79

Appendix B: List of seismic stations in the Western Carpathians—*Permanent Geodynamical Monitoring of Poland—stage I* project (in alphabetical order)

No.	Station code	Place	Registration from	Latitude [°N]	Longitude [°E]
1	BOBR	Bóbrka	2013.11.20	49.62	21.71
2	CICH	Ciche	2013.08.27	49.39	19.87
3	DURS	Dursztyn	2013.08.27	49.41	20.19
4	DZIA	Dzianisz	2013.08.27	49.32	19.84
5	FALS	Falsztyn	2013.08.27	49.42	20.28
6	HARK	Harklowa	2013.08.27	49.48	20.18
7	KACW	Kacwin	2013.08.27	49.36	20.30
8	MANI	Maniowy	2013.08.27	49.45	20.28
9	POTK	Potok	2013.08.27	49.36	20.16
10	ROPK	Ropki	2013.11.19	49.46	20.16
11	SIER	Sierockie	2013.08.27	49.36	19.96
12	STBY	Stare Bystre	2013.08.27	49.43	19.94
13	STOD	Kluszkowce	2013.08.27	49.44	20.29
14	TRBS	Trybsz	2013.08.27	49.42	20.12

References

Cao AM, Gao SS (2002) Temporal variations of seismic b-values beneath northeastern Japan Island arc. Geophys Res Lett 29:9

Draber D, Guterch B, Lewandowska-Marciniak H (eds) (1998) Seismological bulletin 1989–1990, local earthquakes recorded by Polish seismological stations. Publ Inst Geophys Pol Acad Sci B-17(283)

Draber D, Guterch B, Lewandowska-Marciniak H, Niewiadomski J (eds) (2001) Seismological bulletin 1998, local earthquakes recorded by Polish seismological stations. Publ Inst Geophys Pol Acad Sci B-26(338)

Draber D, Guterch B, Lewandowska-Marciniak H (eds) (2002) Seismological bulletin 1995–1997, local earthquakes recorded by Polish seismological stations. Publ Inst Geophys Pol Acad Sci B-24(336)

Draber D, Guterch B, Lewandowska-Marciniak H, Niewiadomski J (eds) (2003a) Seismological bulletin 2000, local earthquakes recorded by Polish seismological stations. Publ Inst Geophys Pol Acad Sci B-32(362)

Draber D, Guterch B, Lewandowska-Marciniak H, Niewiadomski J (eds) (2003b) Seismological bulletin 2001, local earthquakes recorded by Polish seismic stations. Publs Inst Geophys Pol Acad Sci B-33(366)

Draber D, Niewiadomski J, Zuchniak M (eds) (2004) Seismological bulletin 2002, local earthquakes recorded by Polish seismological stations. Publ Inst Geophys Pol Acad Sci B-36(375)

Gutenberg B, Richter CF (1944) Frequency of earthquakes in California. Bull Seismol Soc Am 34:184–188

Guterch B (2006) Seismic events in the Orava–Nowy Targ Basin, Western Carpathians, November 30, 2004–December 2005. Acta Geodyn Geomater 3(143):58–95

Guterch B (ed) (2007) Seismological bulletin 2004, local earthquakes recorded by Polish seismological stations. Publs Inst Geophys Pol Acad Sci B-40(397)

Guterch B (2009) Sejsmiczność Polski w świetle danych historycznych (seismicity in Poland in the light of historical records). Przegląd Geologiczny 57(6):513–520 (in Polish)

Guterch B (2014) Seismicity in Poland: updated seismic catalog (this issue)

Guterch B, Lewandowska-Marciniak H, Labak P, Niewiadomski J (2000) Macroseismic data of seismic events in 1992–1993, Krynica region, Western Carpathians. Publ Inst Geophys Pol Acad Sci B-21(316):133–147

Guterch B, Lewandowska-Marciniak H, Niewiadomski J (2005) Earthquakes recorded in Poland along the Pieniny Klippen Belt, Western Carpathians. Acta Geophys Pol 53(1):27–45

Kijko A (2010) Seismic hazard assessment for selected area. File: Hazard Area Documentation. doc (ver. 21 March 2010)

Kijko A, Sellevoll MA (1989) Estimation of earthquake hazard parameters from incomplete data files. Part I. Utilization of extreme and complete catalogs with different threshold magnitudes. Bull Seismol Soc Am 79:645–654

Kijko A, Sellevoll MA (1992) Estimation of earthquake hazard parameters from incomplete data files. Part II. Incorporation of magnitude heterogeneity. Bull Seismol Soc Am 82:120–134

Kijko A, Singh M (2011) Statistical tools for maximum possible earthquake magnitude estimation. Acta Geophys 59:674–700

Kováč M, Bielik M, Hók J, Kováč P, Kronome B, Labák P, Moczo P, Plašienka D, Šefara J, Šujan M (2002) Seismic activity and neotectonic evolution of the Western Carpathians (Slovakia). EGU Stephan Mueller Spec Publ Ser 3:167–184

Labák P, Brouček I (1996) Catalogue of macroseismically observed earthquakes on the territory of Slovakia (version 1996, in Slovak). Geophys Inst Slov Acad Sci, Bratislava 15

Laska W (1902) Die Erdbeben Polens (in German). Mitteilungen der Erdbeben-Kommission der Kaiserlichen Akademie der Wissenschaften 8:1–36

Maj S, Teisseyre KP (2014) On the roots of the institute of geophysics, Polish academy of sciences. In: Bialik R et al (eds) Achievements, history and challenges in geophysics. GeoPlanet GeoPlanet: Earth and Planetary Sciences, Springer International Publishing, Switzerland, pp 27–38

Mignan A, Woessner J (2012) Estimating the magnitude of completeness in earthquake catalogs, community online resource for statistical seismicity analysis. Available at http://www.corssa.org

Pagaczewski J (1972) Catalogue of earthquakes in Poland in 1000–1970 years. Publs Inst Geoph Pol Acad Sci 51:3–36

Shi Y, Bolt BA (1982) The standard error of the magnitude-frequency b-value. Bull Seismol Soc Am 72:1677–1687

Tóth L, Mónus P, Bus Z, Györi E (2008) Seismicity of the Pannonian basin. In: Earthquake monitoring and seismic hazard mitigation in Balkan countries NATO science series IV: earth and environmental sciences, vol 81, pp 99–110

Trojanowski J, Plesiewicz B, Wiszniowski J, Suchcicki J, Tokarz A, (2012) Monitoring zagrożenia sejsmicznego obszaru Polski (Monitoring of seismic hazard in the area of Poland) (in Polish, final report of the project, not published)

Trojanowski J, Plesiewicz B, Wiszniowski J (2015) Seismic monitoring of Poland—description and results of temporary seismic project with mobile seismic network. Acta Geophys 63, doi: 10.2478/s11600-014-0255-0

Wiejacz P, Kotwicki A, Rudziński Ł (2008) Microearthquakes near Grybów, Poland, Western Carphatians. In: EGU General Assembly, Vienna, Austria, 13–18 April 2008. Geophys Res Abstr, vol 10, p EGU2008-A-05278

Wiemer S, Wyss M (2000) Minimum magnitude of complete reporting in earthquake catalogs: examples from Alaska, the Western United States, and Japan. Bull Seismol Soc Am 90:859–869

Woessner J, Wiemer S (2005) Assessing the quality of earthquake catalogues: Estimating the magnitude of completeness and its uncertainty. Bull Seismol Soc Am 95:684–698

Wyss M, Hasegawa A, Wiemer S, Umino N (1999) Quantitative mapping of precursory seismic quiescence before the 1989, M 7.1 off-Sanriku earthquake, Japan. Annali Di Geofisica 42:851–869

Zsíros T (2003) Earthquake activity and hazard in the Carpathian basin I. Acta Geod Geophys Hung 38(3):345–362

Zsíros T (2005) Seismicity of the Western-Carpathians. Acta Geod Geophys Hung 40(3–4):455–467

Chapter 5
Source Materials to the Outer Western Carpathian Earthquake of December 3, 1786

Barbara Guterch, Hanna Lewandowska-Marciniak and Maciej Mazur

Abstract In this chapter we are listing the previously known and newly found documents relating to the Outer Western Carpathian earthquake of December 3, 1786. They have been the basis for reinterpretation of the parameters of this event, unusual in the area, as described in the first chapter of this book. The manuscripts and articles/notes in contemporary journals are left in original languages (Polish, Latin, German, French). We also provide a catalogue of localities where the evidence of the earthquake was available, together with macroseismic descriptions of the recorded phenomena from various sources, translated into English.

Keywords Macroseismic description · Historical earthquakes

5.1 Introduction

This chapter is a supplement providing details of the previously known and newly found documents relating to the Outer Western Carpathian earthquake of December 3, 1786. They have been the basis for reinterpretation of the parameters of this event, being one of the strongest and deepest earthquakes ever recorded in the area, as described in Chap. 1. The manuscripts and articles/notes/mentions in contemporary journals are left in original languages (Polish, Latin, German, French). We also provide a catalogue of localities where the evidence of the earthquake was available, together with descriptions of the event from various sources, translated into English.

B. Guterch (✉) · H. Lewandowska-Marciniak · M. Mazur
Institute of Geophysics, Polish Academy of Sciences, ul. Księcia Janusza 64, 01-452 Warsaw, Poland
e-mail: bguterch@igf.edu.pl

B. Guterch and J. Kozák (eds.), *Studies of Historical Earthquakes in Southern Poland*, GeoPlanet: Earth and Planetary Sciences,
DOI 10.1007/978-3-319-15446-6_5

5.2 Source Materials to the December 3, 1786, Earthquake

5.2.1 *Materials Found in Manuscripts of Ecclesial and Municipal Archives (Original Spelling and Grammar)*

Bielsko-Biała AP

Archiwum Państwowe w Bielsku, Gerichts protocoll 1785-87, fasc.326, fol.12
Am 3 Decemb. wird vermerket dass allhier abermals nachmittags um 5. Uhr eine Erderschütterung, welche be.laufig 1/v Minute gedauert hat, gewesen se. d, wobe. aber im hochsten se. es gedanket, kein Unglück geschehen.

Cieszyn AP

Archiwum Państwowe w Cieszynie, V Abtheilung des Gedenk-Buchs der Stadt Teschen vom Jahre 1767 bis 1822, fasc.74, fol.114
Am 22. August n.J. früh um 7 Uhr wurde die Erde erschüttert, eine Naturerscheinung, die sich vielleicht noch nie in unserer Gegend ereignet hätte, u. an die man gar nicht geglaubt hätte, wenn sie nicht am 27 Februar früh 4 Uhr u. am 1. Advent-Sonntage 1786 gegen 5 Uhr Abends mit größerer Heftigkeit zurück gekehrt wäre. Beide Erdstösse waren so heftig, dass durch ersteren der neu gemauerte hintere Theil des Convictsgebäudes Risse erhielt, u. aufgestellte Ziegeln herabfielen, die Gläser im Speisezimmer stießen gegen einander, u. durch die letzte wurden sogar die Registraturkästen in der Kreiskanzlei umgeworfen, u. ein Stück der Stadtmauer beim Oberthor zusammengerüttelt.

Kęty MO

Muzeum Okręgowe, Oddział w Kętach, Księga Cechu Tkackiego Barchanickiego, fasc.H/697
Uznawszy za rzecz potrzebną podania dalszemu wiekowi do wiadomości przytrafione za czasów naszych okropne czyny a spodziewając się od Czytającego i Słuchającego nabożnego za Dusze moje i współbraci moich do Boga westchnienia, te wam Kochani Successorowie nasi czynię wiadomość:
1785. A=D 22 Augusti o godzinie 7 z rana nigdy tu w Kentach nie praktykowane ani w okoliczach tych tu słuchane dało się uczuć trzęsienie ziemi i szum podziemny, przed którym deszcze obfite prawie od zaczęcia Augusta padały po których słotach dni 4 pogodne i ciepła nastąpiły po tym trzęsieniu cały /?/ w słoty obfity tak, iż trudno było zboża z pól pozbierać.
1786 A=D 27. Febr. o godzinie 4 1/4 z rana drugi raz tu w Kentach i na mil kilkadziesiąt w okolic było strasnieysze i dłuższe trzęsienie ziemi gdyż trwało przez 3 zdrowaś Marya, które wielką trwogą lud przeraziło, w tymże samym Roku zacząwszy a 1– Augusti tak znaczne i częste były deszcze iż Krescencyi zbierać nie dozwalaly aż do 16 Obra gdzie mrozy tęgsze nastąpiwszy, zima się rozpoczęła, dla słot i ustawicznych prawie niepogód zatrzymane na polu zboża śniegiem zasnute zostały a z tych powodów niektórzy gospodarze bardzo mało a drudzy w cale nic na zimę nie siali.
4. Grudnia 1786 r. Nobiles et Spectabiliter Kaspar Proconsul, Wohlman, Watkanowski

Dla uproszenia miłosierdzia Pana Boga o odwrócenie kary i trzęsienia ziemi ma być wotywa w Kościele archiprezbiterialnym w następujący czwartek na którą cechy mają być obesłane aby były z świecami, sumpta ex arario.
1786. D 3. DCbr. o godzinie szóstey wieczór Trzecie trzęsienie tu ziemi było tęższe dłuższe i nayokropnieysze od którego więcey zachoway Nas Boże. Ziemia się nadzwyczaynie trzęsła, ściany się w domach ruszały, powały trzaskały, okna brzęczały, z słabszych szyby wylatywały, dachy się z jedney strony na drugą pochylały, deszczki, gonty zlatywały, kury z grząd spadały, zgoła wielki strach na ziemi już ostatniego ginienia okazywał się. W Ziemi zaś wielki gruch wzruszył się, i tak się widziało jakoby wozów ciężko obładowanych wielki gmin z szelestem szedł, w Kościołach sklepienia padały, mury roztępowały, na Ratuszu Kentkim Zegar z trybu wyskoczył i we dzwon uderzył, w innych zaś mieyscach dęby do połu łomało, w lasach drzewa obalalo, trwało to trzęsienie przez trzy pacierze.

Kraków AKAP-1
Archiwum Prowincji Zakonu O.O. Kapucynow w Krakowie, Chronologia seu Historia Conventus Rosvadoviensis, fasc.AKR.26, T.I, fol.97
1786, 3 Decembris fuit sensibilis terre motus Sandomiriae e circa circum.

Kraków AKAP-2
Archiwum Prowincji Zakonu O.O. Kapucynów w Krakowie, Kronika Klasztoru Krakowskiego, fasc.AKK 95, fol.240
1786, Die 3tia Hora Decembris 5ta pomeridiana fuit hic Cracovia et in Viciniis Terra motus sat notabilis et periculosus ad eo in aliquibus locis ceciderint Camini disrupti sunt muri Eclesiae Augustianorum, Fluvii Vistulae a aqua elevata in manento ad ulnam; diversa signa ruinae in pluribus domibus relicta. Protendebatur hic Terrae motus a Septemtrione ad meridiem in latum extendebatur Petricovia Pulviam, et ultra duravit per haut 5 secundos.

Kraków AKK
Archiwum Kapituły Katedralnej w Krakowie, Acta Actorum Capituli 1776–1787, fasc.25, fol.714–715
Praesents Reverendissimi Domini Custos Venerabili Sierakowski, Michałowski, Boderkiewicz, Wyczałkowski, Woyczyński Datote Canonici, Sabbatho 9 Decembris
Inter principales saeculi moderni eventus maxime animadversions causus primum saeculo praesenti in hoc Regno practicatos motus scilicet graves trementis terrae binis vicibus annocurrenti ora nostra sentire passa est die 27- February hora quarta post mediam noctem accidit huiusmodi motus Cracoviae in aliquibus vicinioribus solummodo Territoriis eiusdem Palatinatus secundariae nempe die 3-tia Mensis Currentis hora 5-ta vesperrina longe vehementior non tantum ipsam Cracoviam et confinia verum etiam partem Regionis notabilem signater Territoria Palatinatum Cracoviensis Sandomiriensis ac etiam in parte Siradiensis Maioris Poloniae tanto tremore per tinuit, ut etiam signa variis in locis in commotus structuris reliquerit ute Casimiriae ad Cracoviam in Ecclesiis SSnis Corporis Canonicorum Lateranensium notabilius S.Catharinae Religiosorum Fratrum Augustianorum

in cuius fornice scissarce latepatentes cum periculo lapsus sunt apertae. Haec omnipotentis manus correptio incusso timore populum christianum humiliando universam Diaecesiam nostram ad fundendes preces indictas supplicationesque duraturas excitavit. Luce etiam posteris notificanda Reverendissimi Domini in actis suis annotari mandarunto.

Kraków AKMK-1

Archiwum Kurii Metropolitalnej w Krakowie, fasc.A.Ep.112, List pasterski prymasa Polski, fol.203

...Elementa same co raz bardziey się przeciw nam wzmagaią, ziemia co raz z wieksza mocą i w większey rozległości w Kraiach naszych się trzęsie, ustawne deszcze i wylewy nie mało szkody w wielu poczyniły mieyscach, inne ogniem uszkodzone smutny tylko obraz wystawa mnożących się ruin w naszey Oyczyźnie. Powietrze iakoby zmieniło pory Roku, ni nam doyzrzałego nie dopuściło zbierać zboża, ni przyszłe nadzieie dostarczaiącym wspierać zasiewem pozwala. Zkąd niedostatek dzisieyszy przyszłym grożący nieurodzaiem, głód i wynikaiące z niego nieszczeŚliwości wcześnie zapowiadaią. A choroby i już tu i ówdzie z niedostatku i niestateczney pory czasów się wzmagaiące, naysmutnieyszych lękać nam się każa skutków...

Kraków AKMK-2

Archiwum Kurii Metropolitalnej w Krakowie, fasc.A.Ep.112, Zalecenie do kościołów farnych i zakonnych, fol.252–253

Die V Ejusdem. Zalecenie do Kościołów Farnych i Zakonnych Krakowskich obojej płci względem odprawiania Wotyw z Supplikacjami ażeby Pan Bóg zachował nas potym od klęsk trzęsienia Ziemi które na dniu trzecim tego miesiąca po piątej godzinie wieczornej w Krakowie i Okolicach uczuć się dało.

Trzęsienie Ziemi, które dnia onegdajszego to jest trzeciego tego Miesiąca w tym Mieście i okolicach tak znacznie uczuć się dało, drugi raz już w tym Roku umysły wszystkich przerażając, gdy na pamięć przywodzi najokropniejsze klęski, które z tej plagi Boskiej odległe od nas Kraje poniosły... powinno nas pociągać na wzór Niniwitów do uczynkow pokutnych, do Ofiar błagalnych, któremi byśmy sobie miłosierdzie Boskie zjednać i zagrożone Nam ukaranie odwrócić mogli. Na ten koniec Urząd Biskupi rozrządza, aby po wszystkich Kościołach w Krakowie i na Przedmieściach odprawione były Wotywy uroczyste przybierając w nich Collectam tempore Terrae motus (: z Wotyw pod Tytułem Proquacumque necessitate:) z wystawieniem Najświętszego Sakramentu w Puszce z Litaniami o WW SS z Supplikacjami zwyczajnemi Święty Boże do których odprawienia dni się niewyznaczają, ale tych wybranie zostawia się Rządcom Kościołów, byle się bez odwłoki odprawiły po wszystkich Kościołach Świeckich i Zakonnych tak Męskich jako i Panieńskich. Wszyscy zaś Kapłani Świeccy i Zakonni przy Mszach Świętych mają przybierać przez cztery Tygodnie wspomnioną kollekte tempore Terrae motus...

A Kaznodzieje i inni Naukę Duchowną w Kościołach dający zagrzewać będą lud wierny do uczynków pokutnych i Modlitw, aby Nas BÓG miłosierny zachował od wszelkich klęsk, i tej z trzęsienia Ziemi słuszną bojaźnią okolice nasze przerażającej. Które Rozporządzenie Urzędu Biskupiego aby swoje dopełnienie nieodwłocznie odebrało, ma być wszystkim Przełożonym Kościołów i Klasztorów tak Męskich jako i Panieńskich w Krakowie i na Przedmieściach okazane. Dan w Krakowie Dnia 5. Grudnia Roku 1786.

Kraków AKMK-3

Archiwum Kurii Metropolitalnej w Krakowie, fasc.Ag.10, Acta Gratiosa, fol.72
(Recommendation about the prayer as in the document Kraków AKMK-2)

Kraków AKMK-4

Archiwum Kurii Metropolitalnej w Krakowie,Raporta ex Cancellaria Oficij Consistorij Sandomiriensi ex Mense Decembri Anni 1786, fasc.Ag.12, fol.386
His praemissis omitti non potest, quod sub die 3-tia Mensis praesentis hic Sandomiria evenit. Nempe hac die celebri ob Dcam. I-mam adventus pro tunc incidentem inter 5-tam et 6-tam horas pomeridianas motus terre per omnem hance Civitatem consurrexit, duravitque unum circiter minutum horae, absque tamen ulla (: notabiliori praesertim:) violentia et validitate quinimo sic lenis et discretus ut non nisi ab ijs qui protunc in quiete et silentio versabant, sentiri et observari potuerit, jam autem serijs ccupati, aut Clamorosius colloquentes nullum illius experti fuere. Hinc etiam nulla post se reliquit vestigia in alijs rebus praeterquam in hominum memoria, cui se per sensus illorum impressus. Variae de illius rei, observatione spargebantur relationes, sed nullae adeo certae, quam quae factae ab omnibus. Infirmis in lecto pro tunc decumdentibus quibus sub eundem motum lecta eorum in altum erigi visa fuere. Quod ipsum egomet infra sciptus proprio experimento didici, qui pro tunc quiescens in curuli (: vulgo na kanapie:) bene mihi praesens observari, quo modo eadem: ut jam hac utar voce curulis sursum mecum elevabat. a Meridie, et declinabatur ad Septemtrionem itaque tribus ita vicibus et simultanee ad 3-tiam eius modi vicem sonum Campanularum sive Campanulae, horologij in turri Collegij Academici existentis velociter et cum tremore editum, animadverti, Unde protunc ad hanc tertiam vicem vel maxime intuli et mini omnino persvasi huncce eventum nihil aliud esse quam commotionem terre.

Kraków AKMK-5

Archiwum Kurii Metropolitalnej w Krakowie, fasc.Ag12, Raport z Oficialiatu Wiślica, fol.388
Die Prima Dominica Adventus quae erat 3 Decembris terrae motum circumferentia Officaliatus Vislicensis uno eodemque fere tempore scilicet quadrante pro hora sexta pomeridiana aut media sexta sensit hac solum differentia quod aliis in locis fortiorem aliis debeliorem.

Kraków AKMK-6

Archiwum Kurii Metropolitalnej w Krakowie, fasc.Ag.12, Raport Miesiąca Grudnia z Dekanatu Kijskiego Roku 1786, fol. 391
Doniesienie Raportu Dekanatu Kijskiego okazuje w całym Dekanacie trzęsienie Ziemi, a lub różnie z różnych Parafii piszą jednakowoż na jeden dzień, i godzinę jedną zdają się, że dnia 3go Grudnia w pierwszą Niedzielę Adwentu o godzinie wieczornej piątej toż nastąpiło. Dla lepszego Objaśnienia Raporty Niektórych Parafii, w których znacznie dało się uczuć Ziemi trzęsienie kładą się, inne zaś nic nadto nie donoszą, w powszechności się donoszą.

Korytnicka Parafia donosi że tak znaczne było w Dworze u JW. Krajczostwa trzęsienie, że przytomni Goście zostawszy w zadziwieniu mówić do siebie nie mogli, prócz jednego który gwałtu zawołał, ato iż bardzo wielkie wzruszenie czuć się dało, tak Pieca jako też łoskot drzwi dźwięk okien.

Pieńczowska Parafia donosi także znaczne Ziemi trzęsienie przed którym szelest dał się słyszeć niemały na kształt jazdy, od którego trzęsienia sygnaturka na kościele dała się słyszeć.

Parafia Gnoińska donoszą że w całej parafii było ziemi trzęsienie wielkie że Dzieci z przypiecków spadały, piece się rysowały wierzchy budynków trzeszczały. Ptactwo Domowe nawet z grzęd spadało.

W wsi Widuchowej Parafii Gnoińskiej tak straszne było że aż wonty na Sadzaw w Toniach wyciętych burzyły się.

Parafia Lipowska, także też donosi o trzęsieniu Ziemi iż w niektórych wioskach bylo silniejsze w niektórych zaś leksze jako i w Plebanii, które wspokojności tylko bedący uważać mogli. Lecz oprócz tego w tejże parafii we wsi Brudzowie w Dworze powtórnie dało się czuć Ziemi trzęsienie. Dnia 9 Dbri o godzinie ósmej przy wschodzie słońca toż za trzema zawodami dało się poznać przez ruszenie się Haczyków u drzwi odrywanie się zamków, uderzanie prętów oderwanych o szyby u okien. Trwało każde wzruszenie Ziemi jak pięć lub sześć słów wymówić można. Z innych parafii także donoszą iż było trzęsienie ziemi leksze tęższe lecz nic do doniesienia znaczniejszego niemasz nad te które się wyżej wyraziły.

Kraków AKMK-7

Archiwum Kurii Metropolitalnej w Krakowie, fasc.Ag.12, Raport z Dekanatu Koprzywnickiego w Miesiącu Grudniu RP. 1786, fol. 393

Dnia trzeciego tego miesiąca około godziny piątej w wieczór rzadko w kraju tym owszem przez dziś żyjących niedoznane nigdy trzęsienie ziemię przeraziło mocno mieszkańców, które w tym Dekanacie koprzywnickim trwało nie więcej jak minutę jedną. Wzruszenie to ziemi wolnemu od wszelkiego zatrudnienia i zabaw, a w spokojności się zachowującemu czuć się dało do zadziwienia, po wszystkich domach i zgromadzeniach w tym czasie najwięcej mówią o tym trzęsieniu, w prostości Ducha parafianie za zagrożenie od Boga sobie to tłumaczą żądając i wymagając po Pasterzach swoich (co nie jest ich władzą:) publicznych a solennych po Kościołach suplikacyi aby tym sposobem Boga zagniewanego błagali. (…) Dan w Skotnikach dni ostatnich miesiąca Grudnia R.86.

Kraków AKMK-8

Archiwum Kurii Metropolitalnej w Krakowie, fasc.Ag.12, Raport Dekanatu Wolbromskiego dnia 31 Grudnia 1786, fol. 395

Z Wszystkich Far Dekanatu doszli mnie doniesienia względem trzęsienia Ziemi, które dnia 3 Grudnia po godzinie piątej podwieczornej uczuć się dało, w Farze mojej równie cała Parafia tego okropnego Ziemi skutku doświadczyła, lubo mniej albo więcej podług miejsc wyżej lub niżej położonych.

- Więcej się w Dekanacie nic nie zdarzyło, czego by wiadomość interesować mogła"

Kraków AKMK-9

Archiwum Kurii Metropolitalnej w Krakowie, fasc.Ag.12, (Raport Z Parafii Kromołów), fol.397

Dnia 3 Grudnia po godzinie piątej wieczór, dało się tu uczuć znaczne trzęsienie Ziemi, które ja sam dosyć czuło uważałem, kiedy się ściany całego budynku drewnianego zatrzęsły. Kieliszki w kredensie stojące dzwoniły. Drzwi się same od tego wzruszenia otwierały, a podemną przy stoliku na ten czas stojącym, nogi się dobrze jak na bagnie zachwiały. W miasteczku i po wsiach do parafii należących podobne skutki uczuć się dały. Woda z Rurnic w Gorzelni na ziemię pluskała, Garki w szafach stojące, kury na grzędach siedzące na ziemię spadały, a huk wielki pod ziemią dał się słyszeć. Atoli przecież dzięki Panu Bogu, że takowe Ziemi trzęsienie, żadnej nie uczyniło winy, fatalnej nie przyniosło szkody, w całej tutejszej parafii. Ks. Jan Rucki

Kraków AKMK-10

Archiwum Kurii Metropolitalnej w Krakowie, fasc.Ag.12, Rapport z Parafyi Kościoła Mrzygłodzkiego Mil dziesięć od Krakowa to Miasto położone, fol.399

Dnia 3 Miesiąca Grudnia Roku 1786 że: dnia tego, o godzinie piątej y kwadrans wieczór, dało się uczuć trzęsienie Ziemi, które mogło trzymać nad minutę dwie nie więcej. Nayprzód okrutny szum bal, zasię wiatr najgwałtowniejszy, a po tym kiedy największe bryki słyszeliśmy jadące, w tym wzruszyły stoły, krzesła, a na kredensie szklanki y kieliszki, w tym ściany y dachy poczęły trzeszczeć, w niektórych domach miejskich poczęły garnki y sprzęt gospodarski ruszać się, drzwi w izbach y chlewach otwierać się, dosyć że ludzie jedni na dwór poczęli uciekać a inni nie wiedzieli co czynić. Po innych wsiach do tej parafii należących, a ku Siewierzowi wyciągających się, słyszeli ludzie huki podziemne, jakoby z armat strzelano, wszystkie domy y chałupy wzruszone były. Drzwi się otwierały, szyby pękały, garnki y inne naczynia z półek spadały, rozumieli Ludzie że się ziemia z niemi zapada.

Kraków AKMK-11

Archiwum Kurii Metropolitalnej w Krakowie, fasc.Ag.12, Z Parafii Potockiey Dekanatu Lelowskiego, fol.401

Z Parafii Potockiey Dekanatu Lelowskiego Woiewóctwa Krakowskiego wiadomo czynie że 3 Gentis circa horam quintam vespertinam po całey Parafii było uczucie trzęsienia ziemi, ia prawda nieuważałem, bo też nie było zarówne po wszystkich mieyscach, y w samym Potoku. Kiedy w iednych domach mało zważali, a w drugich lepiey doświadczali, a osobliwie w dwóch chałupach, aż kury z grząd spadowały, w Janowie zaś Miasteczku trafiło się, że się aż drzwi do izb poczyły otwierać, iednak nieszkodliwie, ale straszliwie. Słyszałem, że po niektórych polach mieyscami ziemia daie sie widziec popadana. (...)
attestor Datt in Plebanatu Potocensi Die 13 Mensis Decembris Anno Domini 1786

Kraków AKMK-12

Archiwum Kurii Metropolitalnej w Krakowie, fasc.Ag.12, Rapport Dekanatu Nowogórskiego w Miesiącu Styczniu Roku 1787, fol.465

Z okazji trzęsienia ziemi w miesiącu Grudniu Roku zeszłego, wszczęły się w Parafii tej choroby, jednak nie fatalne.

Kraków AKMK-13

Archiwum Kurii Metropolitalnej w Krakowie, fasc.Ag.12, (Raport z Parafii Trzebinia), fol.469

W Tej Parafii z okazji trzęsienia Ziemi i innych okoliczności pogroziło Boskich strwożony i do Nabożeństwa zachęcony którym ledwie można wystarczyć ile w spowiedzi.

Kraków AP-1

Archiwum Państwowe w Krakowie, Archiwum O.O. Augustianow, fasc.22, Kronika Klasztoru, fol.109

1786...tertia Decembris accidit quadrante post horam V. vespartinam cum maximo Omnium terrore et luctu, hac de causa motus terrae fornix Ecclesiae nostrae supra Presbyterium in sex distinctis locis est disrupta, cum maximo periculo, transeuntians, quae citissime fuerat reparata...Prioris Conventus pro tunc existentis.

Kraków AP-3

Archiwum Państwowe w Krakowie, Teki A. Schneidra, fasc. 1623 (trzęsienie Ziemi w 1786 – raporty urzędów powiatowych w Myślenicach, Nowym Sączu i Tarnowie)

An Ein Hochlöbl: k:k Landes Gubernium Das Sandezer Kreis-Amt macht die Anzeige von einer für gewesten Erd-Erschütterung.

Hochlöbliches Landes Gubernium,
Die außerordentliche den Kreis betreffende Begebenheiten findet man anzuzeigen des Dinstes zu seyn. Hinzu eignet sich die am 3-ten dieses nach Mittags zwischen 5 und 6 Uhr und dem Vermuthen nach von Westen her im ganzen Kreise empfundene Erd-Erschütterung. Hiervon macht man dann die berichtliche Anzeige mit dem Zusatz, das selbe keine sichtbare Beschädigungen veruhrsachet habe.
Neu-Sandecz d 6 Decemb. 1786.
(undecipherable signature)

Hochlöbl. K.K. Landes Gubernium
Gestern, als den 3-ten dieses ist hier in Myslenice ein Erdbeben gewesen, welches 2 Minuten angehalten hat und ziemlich heftig war, da in einigen Häusern Stühle, Bänke und andere Mobilien zu Boden geworfen wurden; Wann aus den anderen Gegenden dieses Kreyses hierüber nähere Anzeigen eingehen sollten, wird man unvergessen seyn, solche in Gehorsam nachzutragen.
Myślenice den 4' D-ber 1786.
(undecipherable signature)

An ein Wohllöbliches KK Landes Gubernium
Bericht des Tarnower Kramts wegen des Erdbebens welches sich den 3-ten D-ber in Tarnow verspüren lies.

Hochlöbliches Landes Gubernium!
Den 3-ten Christmonats Nachmittags um 5 Uhr hat sich in der ganzen hiesigen Gegend ein ziemlich starkes Erdbeben verspühren lassen, vom platten Lande hat man noch keine Nachricht, dass irgendwo einen Schaden verursachet hätte. In der Stadt Tarnow hat es in 8 Häusern einige Risse in deren Wänden verursachet, wo man unverzüglich durch den Kr. Ingenieur eine genaue Untersuchung derenselben hat vornehmen lassen; in 5 Häusern sind diese Risse ganz ohne aller Bedeutung, in dreyen hingegen dürften sie nachtheilig werden, bey selben hat man nach Umständen die nöthige Einleitung getroffen, um aller Gefahr vorzubeügen. Tarnów d 5=ten Decemb 786.
(undecipherable signature)

Bericht des galiz. Landes Gubernium
Die Berichte in betref Hinlandes verspürten Erdbebens werden gehorsamst unterlegt.
Lemberg den 12. Decemb 786
(Hofkanzley)
Vermög den hier gehorsamst anverwahrten Berichten ist nicht nur in Myslenize, sondern auch in Tarnow und im Sandecer Kreise am 3 dieses ein Erdbeben verspührt und in Tarnow selbst 3 Häuser ziemlich beschädiget worden, wobey aber das Kr:Amt die nöt His praemissis omitti non potest, quod sub die 3-tia Mensis praesentis hic Sandomiria evenit. Nempe hac die celebri ob Dcam. I-mam adventus pro tunc incidentem inter 5-tam et 6-tam horas pomeridianas motus terre per omnem hance Civitatem consurrexit, duravitque unum circiter minutum horae, absque tamen ulla (: notabiliori praesertim:) violentia et validitate quinimo sic lenis et discretus ut non nisi ab ijs qui protunc in quiete et silentio versabant, sentiri et observari potuerit, jam autem serijs ccupati, aut Clamorosius colloquentes nullum illius experti fuere. Hinc etiam nulla post se reliquit vestigia in alijs rebus praeterquam in hominum memoria, cui se per sensus illorum impressus. Variae de illius rei, observatione spargebantur relationes, sed nullae adeo certae, quam quae factae ab omnibus. Infirmis in lecto pro tunc decumdentibus quibus sub eundem motum lecta eorum in altum erigi visa fuere. Quod ipsum egomet infra sciptus proprio experimento didici, qui pro tunc quiescens in curuli (: vulgo na kanapie:) bene mihi praesens observari, quo modo eadem: ut jam hac utar voce curulis sursum mecum elevabat. a Meridie, et declinabatur ad Septemtrionem itaque tribus ita vicibus et simultanee ad 3-tiam eius modi vicem sonum Campanularum sive Campanulae, horologij in turri Collegij Academici existentis velociter et cum tremore editum, animadverti, Unde protunc ad hanc tertiam vicem vel maxime intuli et mini omnino persvasi huncce eventum nihil aliud esse quam commotionem terre.

Wir sinnen daher nicht diese Nachrichten zur höchsten Wissenschaft zu unterlegen. Lemberg

Zur Sizung den 11-ten Janner in 787.
Hof Kanzley Dekret
Demselben wird auf seinen Bericht vom 12-ten d.M. dessen sämtliche Beilagen hier zurückfolgen, erwidert, dass die eingesendeten Nachrichten von dem gewesenen Erdbeben in Mislenice, Tarnow und Sandezer Kreise zur Wissenschaft dienen.
Wien d 29 Dec: 786.
(undecipherable signature)

Kraków AP-4

Archiwum Państwowe w Krakowie, Księga radziecka miasta Kazimierza, fasc. K-329, fol.541

Dnia 7 Grudnia 1786 r.... Na wotywy Cechy gdy z światłem asystują, niektórzy nieprzystojnie ubrani przychodzą, jako i na onegdajszej wotywie, od trzęsienia ziemi odprawianej widzieć się dało, a przeto zaleca się aby przystojnie ubrani do asystowania wotywom przychodzili.

Kraków AREF

Archiwum Prowincji Zakonu O.O. Reformatów w Krakowie, Kronika Klasztoru Krakowskiego, fasc. II, fol.203

Anno Domini 1786 Die 3 Decembris post horam 5tam pomeridianam factus est notabilis terra motus Cracovia et circa Cracoviam qui brevissimo tempore duravit quasi per medium Ave Maria, observavit tota Communitas existens in Refectario in meditatione et audiens Commotionem Mensarum, hoc ipsum observarunt et alis fratres oficialis in Cellis cacotentes. Imo omnes incola Cracovienses sequenti Die implorando Divinam Misericordiam Cantavimus Missam de S.Barbara post Missam recitavimas Litanias de Nomine Jesu cum Gratione Urbani VIII ante oculos tuos Domine et Consuetis Suffragiis.

Kraków-Mogiła AOC

Archiwum Opactwa O.O. Cystersów, Notatnik zdarzeń kleryka mogilskiego 1785-1789, fasc.11, fol.47

Die 3 Decembris scilicet Dominica primae adventus Anno 1786 Tempore Coene terra motus validus. Omnes timores percusit, praecipue autem Dominum Josepfum Szufranowicz proffesum Andreoviensem pro tunc lectorem mense qui pre pavore Cathedram reliquent fugam capessit. Tanta erat vis, tanta matio ut et Campanulla penes portum collegi pendens sonos ederet, et Imagines parietibus affixe vidente subito ruerunt.

Tarnów AD

Archiwum Diecezjalne w Tarnowie, Notificationes Anno 1781–1789

Florianus amndus Janowski z Bożej i Apostolskiej łaskawości Biskup Tarnowski Jego Cesarskiej Król. Apostolskiej Mości Konsyliarz. Wszystkim duchownym świeckim i zakonnym wiernym Chrystusowi zdrowie i Pasterskie Błogosławieństwo. Abyśmy Boga Naszego zapalczywość, która przez różne ukazania a świeżo 3-go Decembris trzęsieniem ziemi dla poprawy naszej ukazane przebłagali i do miłosierdzia pobudzili, zarządzamy publiczne modlitwy czyli suplikacje, które w kościele naszym Katedralnym i Kolegiatach następującym porządkiem mieć odprawione chcemy z rana Msza z wystawieniem in Pixide Najświętszego Sakramentu o salutaris Hostia, a na Mszy przez Święta prima et secunda clasis ma być kollekta pro avertedo terra motu Omnipotens sempiterne Deus qui facio terra tremere, a po Mszy litania do WWSS w polskim języku z kollektami od powietrza głodu ognia i wojny i trzęsienia ziemi a na koncu Fiant Domine cum benedictione consueta, po południu o godzinie piątej exposito Sanctissimo…w Tuchowie 5 grudnia 1786.

Tarnów MO

Muzeum Okręgowe w Tarnowie, Acta Advocatalia et Scabinalia Tarnoviensia, fasc. MT-H496, fol.44

Działo się w Tarnowie dnia 4 Decembris 1786. Do publicznej każdemu daje się wiadomości, że dnia 3 Decembris 1786 o godzinie 5 po południu minuty jedynastej w mieście Tarnowie okropne trzęsienie Ziemi przez jedną minutę trwało, przez co niejednemu obywatelowi kamienice porysowały się i uderzenie serca w dzwonki Katarzyny i w kościele Ojców Bernardynów podwakrotnie słyszeć się dało, dlaczego wszystkich ten okropny przypadek czytających obliguje się ażeby westchnienie swoje od powtórnego takowego przypadku na Miasto Tarnów uniknienia do Najwyższego uczynili.

Uście Solne

Kościół Parafialny w Uściu Solnym, Liber Copulatorum et Liber Babtisatorum (od roku 1775), fol.49

Roku 1786 Dnia 3 Grudnia godziny piątej wieczorowej po zachodzie słońca. Ziemia się trzęsła tu w tutejszym miasteczku tak dalece że po niektórych Domach zamki u drzwi puszczały, drzwi się otwierały, dachy trzeszczały i ściany jakoby się waliły, okna, piece widocznie się ruszały tak że wielkiego strachu bojaźni płaczu hałasu pomiędzy ludźmi narobił ten przypadek tu niezwyczajny, lubo przez bardzo krótki bo tylko przez którąś minutę trwający, że się drugim i poznać niedał, toż się działo i tu po wszystkiej okolicy w Barczkowie, Popędzynie, Wrzepi, Bratucicach, Bogucicach, Cerekwi, Bieńkowicach etc. co widocznie wszyscy to ludzie i starzy i mali uczuwali, w niektórych jednak miejscach jako zeznawali ludzie dłużej to trzęsienie trwało, w Jaxicach bowiem jako powiadali przez trzy kwadranse po trzy razy się powtarzając, w Bochni się na pułgodziny przed zachodem słońca zaczeła i indziej których tu miejsc wszystkich nie wypisuję, o tutejszym tylko czyniąc wzmiankę przypadku.

Wrocław BOS

Biblioteka im. Ossolińskich, fasc.11634.II, (Raport pułkownika Dawida Grammlicha do generala Jozefa Wodzickiego), fol.985–986

...Mam honor Wielmożnemu Panu Dobrodziejowi donieść, iż u nas dnia 3 praesenti zaraz o godzinie 5 po południu bardzo mocno po całym Krakowie dało się uczuć trzęsienie ziemi z hukiem, którego i ja sam doświadczyłem gdyż siedząc blisko ściany, nagle z stołkiem odepchniętym zostałem, gdzie zdało mi się iż wszystkie mury się obalają, jednakże nie słychać ażeby się gdzie jaka szkoda stała.
Jestem z należytym poważaniem Wielmożnego Pana Dobrodzieja najniższy sługa
Dawid Grammlich Pułkownik, Z Krakowa 6 Decembri 1786.

5.2.2 Materials from Contemporary Journals (Original Spelling and Grammar)

BRÜNNER ZEITUNG, 1786
No 99 p. 787

Tarnov in Galizien, vom 4 December.
Gestern Abends gegen 5 Uhr verspürten wir althier eine Erderschütterung, wodurch aber zum Glück weiter kein Schaden angerichtet wurde

No. 101 pp. 799–800
Plesse vom 6 December
Sonntagst, den 3 diesen Abend 3 Viertel auf 5 Uhr verspürten wir hier eine ziemlich starke Erderschütterung, so dass darüber die meisten Leute aus den Häusern liefen wie denn wirklich einige Gebäude Risse bekamen, und die Stubenöfen dergestalt zerrättet wurden, dass man die Zimmer am folgenden Morgen, wegen des eindringenden Rauches, gar nicht heizen konnte. Hühner und Vögel wurden von ihren Sitzen herunter geworfen, und dies geschah an manchen Örtern und mit den Gläsern. Leute die sich um eben diese Zeit auf dem Felde befanden, versichern übereinstimmend, dass es gewesen sey, als wenn sie mit der Erde hinunter sinken sollten, und dann wieder in die Höhe geworfen würden.

Aus Krakau ist ebenfalls schon die Nachricht von diesem Erdbeben eingegangen, wo es sich aber etwas später nämlich um ein Viertel nach 5 Uhr ereignete, und nach dem Zeugnisse des Thürmwächters auch stärker war als sonst; denn bei der heftigen Bewegung des Thurms schlug der Klöppel zweimal an. Gott Lob! So viel man weiß, so lief doch alles noch ohne Schaden ab.

Hiervon meldet ein Schreiben aus Tarnowitz noch Folgendes: „Den 3 dieses, Abends nach 5 Uhr als ich eben bei einem meiner Freunde war, spürten wir hier eine Erderschütterung. Ich war mit ihm allein auf seiner Stube, und auf einmal sing der Stuhl, M m m m m auf dem ich saß an, sich unter mir so zu bewegen, dass es nicht nur empfunden, sondern auch gesehen werden sonnte. Ich fragte den mir gegenüber sitzende Freund: Fühlen Sie nichts? Er sprang auf, und sagte: Lassen Sie uns ins Freie gehen. Die Wände fingen nun eben an zu wanken, die Fenster zu klirren und die Decke über uns, so wie der Boden unter uns, sich sichtbar zu bewegen; eine Menge Staub erfüllte das Zimmer, und wir eilten der Thüre zu, die ich aber nicht so gleich öffnen konnte, denn sie war durch die noch fortwährende Bewegung geklemmt, und das war unser Glück. Während dieser Zeit nämlich fiel ein Ziegel und mehr Schutt herunter gerade auf die Treppe, wo wir ungefähr gewesen sein würden, wenn die Thür sogleich aufgegangen wäre. Wir eilten ins Freie zu kommen, und fanden auf dem Markte schon mehrere Leute die ihre Wohnungen verlassen hatten. Ein jeder erzählte was er erfahren gehört und gesehen hatte. Das Wetter war sehr stille und trübe; aber gar nicht kalt. Der Wind stand aus Nordost, und die Stöße kamen aus Südwest folglich dem Winde gerade entgegen. Das Beben war ungleich stärker, als jenes vom 27 Februar dieses Jahres; denn es ist fast kein Haus um Markte, welches nicht etwas gelitten hätte u. s. w.“

Aus Ungarn wird eben dass berichtet: Man spürte das Erdbeben in Kaschau, in Kirchdorf und Iglo ober Neudorf an besagtem Tage fast am eben die Zeit wie in Plesse. In Iglo soll es sehr heftig gewesen sein, denn die Thurmglocken wären in Bewegung gekommen, und die Schwendel hätten einigemal angeschlagen, so dass die Einwohner Feuerlärmen zu hören glaubten.

No. 102, p. 811
Brünn, den 22 Decemb.
Das Erdbeben am 3 dieses, welches sich über einige Gegenden von Schlesien, Pohlen und Ungarn erstreckte, wurde auch in Mähren, an der Gränze verspürt. Nach einem Berichte aus Littau war es daselbst so stark, dass unter andern einige Gewölbe große Risse bekamen.

No. 103, p. 817–818
Lemberg, den 15 Decemb.
In Briefen aus Tarnow wird das Erdbeben, welches sich daselbst am 3 dieses ereignete, als sehr heftig beschrieben. Verschiedene Sachen wurden aus ihrer Lage geworfen, und die Einwohner liefen von Schrecken, unter einem jammernden Geschrei, alle aus ihren Häusern, indem sie befürchteten, dass alles über einem Hausen zusammen stürzen würde. Bei dieser Erschütterung hatte die dasige Rathausglocke einige mal von selbst angeschlagen. Zum Glücke muss aber alles ohne Schaden abgelaufen seyn, weil diese Berichte nichts davon erwähnen.
Ungeachtet besagtes Erdbeben in unserer Hauptstadt kaum merklich war, so wurde es dennoch auf Zimmern, wo alles stille war, von verschiedenen Leuten verspürt; und wenn man den aus allen Gegenden des Landes eingehenden einstimmigen Nachrichten glauben darf, so ist nichts gewisser, als dass sich diese Erschütterung durch ganz Galizien verbreitet habe, mit dem Unterschiede, dass die Erde in verschiedenen Ortschaften sich mehr oder weniger bewegte. Ein Schreiben aus Neumark enthält Folgendes: 'Das am 3 dieses hin und wieder verspürte Erdbeben hat unser Neumark in einem höheren Grade getroffen, als viele andere Gegenden. An besagten Tage, des Abends zwischen 5 und 6 Uhr, hörte man auf einmal ein fürchterliches unterirdisches Brausen. Nachdem dieses einige Minuten angehalten hatte, erfolgten gleich nach einander 2 Stöße. Es setzte ungefähr 5 Minuten aus, und es kam der dritte welcher die Einwohner in Todesangst versetzte. Er war so heftig, dass die Felsen auf dem Karpatischen Gebirge, wie man nachher bemerkte, auf einen halben, ja ganzen Schuh breit, aus einander gesprengt wurden. Hier in Neumark selbst fielen fast in allen Häusern die Öfen über den Haufen, die einzigen zwei von Stein gebauten Häuser, die da sind, bekamen von der Grundfeste bis an das Dach solche Spalten und Risse, dass man einen Mannsdaumen hinein legen kann, die hölzernen Häuser, die in der Grundlage nicht genug befestiget waren, wurden durch das Erdbeben von ihrer Stelle gerückt, und völlig verdreht, so dass die armen Leute zu thun haben werden, bis sie ihre Wohnungen wieder in Ordnung bringen'

No. 104, pp. 825–826

Warschau, den 13 Dezember:
Die Nachrichten von dem Erdbeben in diesem Königreiche sind sehr übereinstimmend in Ansehung der Zeit, und in diesem Jahre sind bereits zwei bemerket worden wovon sich das erstere am 27 Feb dieses Jahres eräugnete. Nach den Bemerkungen unserer Geschichtschreiber ist das letztere Erdbeben seit dem Jahre 1000 das zehnte in diesem Lande.

Es sind dievon, nämlich von der Erschütterung am 3 dieses, ungefähr fünf Minuten nach 5 Uhr Abends, verschiedene Berichte eingegangen. In einigen Orten hörte man ein unterirdisches Getöse, als wenn Wagen auf der Strasse führen. Zu Kazimierz, welches der Stadt Krakau nach Mitternacht liegt, wurden alle Häuser bewegt, die Dächer krachten, die Stühle und alles, was in Zimmern war, bewegte sich, die Thüren der Stuben und Keller sprangen auf, die Uhren schlugen vielmals an, und die Soldaten verließen ihre Wachstube, weil dieselbe einzufallen drohte. In der dasigen Marienkirche war alles in Bewegung; die Geistlichen, die sich in ihrem Speisesaale befanden, flüchteten ins Ferne; während dieser Zeit bewegte sich die Weichsel in dortiger Gegend mit aussehrordentlicher Gewalt. Dieß that der Fluss auch dort bei Krakau.

In Bochnia fielen Oefen ein, und das Feuer ward in den Stuben herum geworfen. In vielen Zimmern sind die Balken aus einander gegangen. In Sendomir ist bei den Reformaten das Kloster gesprungen, Gläser und alles, was auf Tischen lag, herunter gefallen. Eben so stark war das Beben auch in Olkus. In Krzesowicza hat die Kirche Schaden gelitten, und die Glocken haben sehr heftig angeschlagen. In Garnka, 3 Meilen von Czenstochau an der Warte gelegen, sind viele Wände geborsten, und in Tyniec ist es gleichfalls sehr heftig gewesen. Alles dieses und noch mehr wird in einem Schreiben aus Krakau berichtet, mit dem Beisasse, dass die Gegenden des Weichselstromes es am stärksten empfunden haben, und dass es dennoch im Karpatischen Gebirge noch heftiger gewesen sey, als anderwärts.

GAZETA WARSZAWSKA

No. 98, December 09, 1786, supplement

Z Warszawy d.9.Grudnia. Mamy z Krakowa wiadomość, że tam w niedzielę przeszłą o godzinie piątej i pięciu minutach po południu, w całym Mieście i jego okolicach dało się czuć trzęsienie ziemi, przez kilka minut trwające, dłuższe i mocniejsze, jako to, którego tam w miesiącu Lutym byli doświadczyli; bez żadnej jednak ruiny znacznej, prócz niektórych niewielkich rys. Drżenie ścian, drzwi, okien, sufitów, we wszystkich domach i gmachach, z początku wolniejsze, coraz się bardziej natężało, a ostatnie wzruszenie było dosyć mocne i gwałtowne. Temu to podobno trzaskowi powszechnemu domów i gmachów przypisać należy szelest i szum, który się podczas tamtego trzęsienia ziemi dał słyszeć po wielu domach. Wysokość Barometru tego dnia w wieczór była 27 calów, Termometr stał w punkcie lodu. Czas, tak przed trzęsieniem, jak i po trzęsieniu, był cichy i pochmurny. Wiadomości odebrane z Pińczowa, Wawrzeńczyc, Nadzowa, i z wielu innych miejsc Województwa Kraków i Sendomir zapewniają, że prawie w tym czasie, obydwa Województwa, czuły skutki tego trzęsienia ziemi.

Z Piotrkowa piszą, że w tym mieście, tegoż wieczora i czasu, podobne trzęsienie uczuć się dało; ale i tam Bogu dzięki! Bez widocznej szkody.

No. 99, December 13, 1786

Z Warszawy dnia 13. Grudnia. Jaką wiadomość o Trzęsieniach Ziemi w Polszcze przytrafionych, dawne Kraju naszego Dzieje nam podały, opisaliśmy to w tegorocznej

Gazecie pod dniem 8. Marca przyłączając tamże ostrzeżenie moralne. Teraz kładniemy uwagę, iż od pierwszego w Roku 1000. zdarzonego i w kronikach opisanego Trzęsienia, upłynęło lat 16. Od drugiego do trzeciego, lat 184. Od trzeciego do czwartego, lat 57. Od czwartego do piątego, lat 46. Od piątego do szóstego, lat 45. Od szóstego do siódmego, lat 10. W tych zaś czasach, w przeciągu szesnasta nie spełna Miesięcy, trzy razy Trzęsienia Polskę naszą nawiedziły, to jest dnia 22. Sierpnia Roku przeszłego, d. 27. Lutego Roku bieżącego, i dnia 3. Grudnia teraźniejszego. Okoliczności okropne tego ostatniego Trzęsienia, z Krakowa przysłane, kładniemy tu, przyłączeniem tamże wiadomości i z miejsc innych: Już to drugi raz w tym Roku, nawiedziło nas Trzęsienie Ziemi, mocniejsze i straszniejsze nierównie tą drugą razą, aniżeli owe pierwsze dnia 27. Lutego przytrafione. W dzień S. Franciszka Xawiera (była to Niedziela Pierwsza Adwentu) to jest dnia 3. teraźniejszego Miesiąca Grudnia, o pięciu minutach po godzinie piątej wieczornej, dał się słyszeć po wszystkich tutejszych Krakowskich Ulicach wielki szelest, jak żeby biegnących z impetem po bruku karet. Zaraz czuć się dało Trzęsienie Ziemi, poczynając od Kazimierza na północ, a to przez wahanie czyli kołysanie (które jest gorsze i niebezpieczniejsze) nie zaś przez perpendykularne podrzucanie czyli podskakiwanie. Z większą wagą były nachylania się ku północy, a niżeli odchylania się nazad ku południowi. Zaczęły zatym wszystkie domy wahać się, jak łódź na wodzie. Razem też poczęły trzaskać dachy, belki i wiązania. Wszystkie drzwi kołatały, wszystkie okna dzwoniły. Szafy i same ściany, widocznie nachylały się. Stołki i krzesła pod siedzącymi, od południa ku północy usuwały się. Podłogi i sklepienia pod nogami, zdawały się uginać. Na Wieży Zegarowej, młoty tegoż Zegara, tak wielki szelest czyniły, że nie tylko szyldwach, ale i inni żołnierze, z Kordygardy uciekli. Także na Wieży Panny Maryi, usłyszawszy Trębacz wielki łoskot, rozumiał że złodziej gwałtem do niego się dobija, przeto wyskoczył i przeciwko niemu z siekierą; lecz gdy postrzegł, że dzwon i z wiązaniem sam rusza się, przerażony tym widokiem, z wieży na pół martwy uciekł. Niektórzy Zakonnicy, i w Refektarzu podówczas znajdujący się, przerażeni strachem, od tegoż swego Refektarza odbiegli. Pod tenże czas, Wisła straszliwie się wzburzyła i na łokieć podniosła się, tak dalece, że przechodzący wtenczas ludzie przez most na tejże rzece pływający, rozumieli że most pod niemi tonie. W Bochni tak dalece wielkie Trzęsienie było, że przez rozstąpione w piecach kachle, widzieć można było ogień palący się; a jak tylko Trzęsienie ustało, znowu się kachle doskonale do kupy zeszły się. Po niektórych izbach drzwi się pootwierały. Na Zamku, i w innych niektórych miejscach, Dzwony Zegarowe, same się kilkakrotnie odezwały. W Sendomirzu, i w Klasztorze JJ. XX. Reformatów, szafki przybite do ścian, znacznie poodsuwały się; po domach Jurystów, papiery z stołów pospadały; na wielu miejscach, szklanki i kieliszki powywracały się; Wisła też znacznie się burzyła. W Olkuszu, w kilku sekundach, trzykroć tak mocno ziemia się strzęsła, że krzesła i stołki po izbach bujały, a ludzie od strachu na ziemię z krzykiem padali. W Krzesowicach, tak był Kościół ztrząśniony, że dzwony na wieży wiszącego brząkanie, słyszane było. We wsi Wielkanocy (około 5 mil od Krakowa) gdy pewny liczył wypłaconą sobie Summę, za przypadłym tymże Trzęsieniem, pieniądze z rąk mu wyskakiwać zaczęły, i żadną miarą do kupy zebrać je nie mógł, ale po izbie ze stołu spadając rozsypały się. Po niektórych miejscach, lustra i zwierciadła z ścian pospadały. W Garnku (o 3. mil od Krakowa nad rzeką Warta leżącym) w pokojach tamecznych, stoły i komody i krzesełka trzęsło, ludzi na pawimencie stojących unosiło, a opierających się o ściany odpychało. W Włoszczowie (o mil 12. od Krakowa) nie tylko w Dworze i w Miasteczku ale i w okolicach, krzesełka pod siedzącymi usuwały się, szklanki i kieliszki tłukły się, drzwi same otwierały się, dzwonki u Zegarów nadzwyczajnie dzwoniły, chusty rozwieszone z sznurów pospadały, ptactwo polowe z pod strzech uciekało, domowe także ptactwo z siedlisk swych pozlatywało, w Kościele Włoszczowskim i w Pałacu poodpadały gzymsy. Na Tyńcu, Kościół miał się porysować. To trzęsienie wielką część kraju kołatało; z Lipowca do Opatowa liczy się mil 15, a w obu miejscach czuć się mocno dało. Inne też miejsca, od Wisły dalekie, równie, a miejscami jeszcze tężej, toż nieszczęście spotkało. Tu w Krakowie i innej szkodzie (oprócz małych rys) nie słyszemy, jak że z jednej Kamienicy kamienny wazon spadł; a na drugiej, podobnyż wazon kręcił się i upadkiem groził, przeto go zdjęto. Ale przestraszenie obywatelów, w tym przeciągu (które do 30 sekund pilnie uważający liczą) tak było wielkie, że niektórym, krew puszczać musiano. Podobało się zatym Miłosierdziu Boskiemu i teraźniejszą drugą już tego Roku razą, tę tak okropną Swą plagę, na samym tylko dla nas przestraszeniu i pogrożeniu, acz coraz już większym, łaskawie jeszcze zakończyć.

No. 100, December 16, 1786

Z Warszawy dnia 16. Grudnia. Dalsze wiadomości o Trzęsieniu Ziemi w naszej Polszcze ostatnią razą przytrafionym, z listów późniejszych Krakowskich wybrane, są następujące. W mieście Krakowie, na wielu miejscach mury porysowały się, osobliwie w Klasztorze JJ. XX. Karmelitów na Piasku. W tymże Klasztorze, pewny Zakonnik, tak ciężko na nogi chorował, że od trzech tygodni, z łóżka ruszyć się mógł; gdy zaś ziemia się zatrzęsła, i ściana w jego Celi nade drzwiami pękła. Odegnał strach od nóg słabość. Tak dalece, że porwawszy się chory z łóżka, sam o swej mocy z Celi uciekł. Tamże w Krakowie. Kościół J.J.X.X. Augustianów na Kazimierzu wielki i wysoki, tak się w sklepieniu In prima navi, czyli w Prezbiterium porysował, że musiano go potym zamknąć dla niebezpieczeństwa. Po wielu miejscach około Krakowa, pospadały gzymsy i krzyże; co się przytrafiło i w Plebanii w Rocimowicach. W Kielcach, w Chęcinach. W Głębowicach, i w całym Xięstwie Zatorskim i Oświęcimskim; a czym bliżej do Krakowa, tym mocniejsze. Dalej nie doszło jak do Małogoszcza; za tym albowiem miejscem, dążąc ku Warszawie czuć się niedało. W Siosławicach (o kilkoro staj od Buska) nowo wymurowane Officyny, tak się porysowały, że mieszkańcy uciekać z nich musieli. W Balicach, przyszedł pewny wieśniak do źródła po wodę; w tym gdy się ziemia zatrzęsła, bliska kamienna wielka skała rozpękła się na dwoje, a z takim hukiem, jak żeby z kilkunastu harmat razem nad uchem wystrzelono; ogłuszony chłop, ledwo przyszedł do siebie. Z Zatora pisano, iż tameczny Doktor JP. Gherii, rodem z Włoch, z Miasta Bolonii, wiedząc z doświadczenia w swoim kraju, iż te straszne przypadki, częstokroć na jednym razie się nie kończą, przez całą noc nie spał, i świadczy, że było drugie Trzęsienie, chociaż lekkie o godz 3. po północy, którego śpiący ludzie nieczuli. Przybył jeden z zagranicy do Krakowa, i o podobnym Trzęsieniu w Opawie przytrafionym, czynił niezawodną relację. W Krakowie, mgły grube i siarką trącące, dotychczas jeszcze panują. Z Piotrkowa podobne Krakowskim mamy wiadomości. Między innemi, w Kolegium JJ. XX. Piarów, widocznie ruszały się piece, drzwi, okna i same ściany. Jeden Zakonnik siedzący na krześle, raptowo z południa na północ usunięty został; a skrzypce, wsparte na stole uderzywszy w stół, i dźwięk wydawszy, znowu na swym stanęły miejscu. W drugiego Celi, krzesło pod siedzącym i stół w tę i ową stronę usuwały się. Trzeci Zakonnik, stojący podówczas, tak mocno wahać się poczoł, że stołu chwytać się musiał, ażeby na ziemię nie padł; przytym postrzegł, że wiszący liniał na murze, oscylacje ruszając się czynił. Czwarty, także stojący, gdy się chwiać począł, bojąc się upadnienia, o mur oparł się, ale go mur od siebie odtrącał. Piątego, przy piecu siedzącego, tenże piec mocno uderzył. W tymże Kollegium kilku uczniów i z Dyrektorem, za powstałym Trzęsieniem, do muru przytuliło się; lecz od tego muru odtrąceni, odskakiwać musieli. Toż samo działo się w Klasztorze JJ. XX. Dominikanów, i po innych miejscach, ale wszędzie Bogu dzięki bez szkody. Z Pułtuska d. 13 Grud: Z rozporządzenia Xiążęcia Jmci Pułtuskiego Szembeka Biskupa Płockiego, Pasterza Naszego, rozpoczęły się tu Publiczne Supplikacje, dla błagania Boskiego Majestatu, z okoliczności Trzęsienia Ziemi, znowu w Polszcze naszej ponowionego.

No. 102, December 23, 1786

Z Warszawy dnia 23. Grudnia. Oprócz dawniej wyrażonych miejsc, ostatnim Trzęsieniem Ziemi skołatanych, liczą się także Kurozwęki, Dobra Ichmość Państwa Wojewodztwa Sendomirskich; w tamecznym albowiem pałacu, drzwi w pokojach pozamykane, raptem się pootwierały, papiery na stolikach szelest czyniły, klucze przy zegarach wiszące brzęczały, boazerie na ścianach łoskot dawały, lustra ruszały się etc.: a to z wielkim przytomnych strachem, jako z rzeczy tam niepraktykowanej. W Racimowicach, podczas tego Trzęsienia, Kopuła blachą obita, z Kościoła spadła. W Racławicach, Zakrystia zapadła się. W Kowali, Dobra Dziedzicznych Xiążęcia Jmci Pułtuskiego Szembeka Biskupa Płockiego, o mil 4. od Krakowa z tej strony Warszawy leżących, przy wielkim jak od biegającej po bruku karety szeleście, tak tameczny Pałac, jak i inne murowane i drewniane domy, mocno się trzęsły; ludzie stojący blisko siebie, od wielkiego ruszania się ziemi, jedni drugich potrącali; w partykularnych domach, garnki na kominach stojące na ziemię pospadały. W Zemboczynie (o ćwierć mili od wyżej wyrażonej Kowali zostającym) tameczny staroświecki, ale dobrze i ozdobnie utrzymany Kościół, tak w ścianach, jako i w suficie, znacznie się porysował.

No. 104, December 30, 1786

Z Wrocławia dnia 7. Grud: Wielu tu w przeszłą niedzielę około 5. wieczornej godziny czuło trzęsienie ziemi bardzo lekkie, tak dalece, iż niektórzy o nim powątpiewają. Ale z innych miejsc Szląska doszłe nowiny donoszą nam o Trzęsieniu ziemi, o którym żadnej nie masz wątpliwości. Tu w Wrocławiu najbardziej było postrzeżon z strony przedmieścia zwanego Oblauische Forstadt, z strony Dohm i Sand. Niektórzy powiadają, iż widzieli zegarki ruszające się na ścianach, i szklanki. W Brieg, Neise, Leobschutz, było to trzęsienie mocniejsze. W Ratibor, na 2. stopy gruby mur rozpadł się, a po rozmaitych domach okna powylatywały. Dalsze o tym trzęsieniu ziemi nowiny, niezaniedbamy donieść.

Z Namlsau d. 4 . Grud: Nasze Miasto wczorajszego wieczora doznało osobliwszego natury przypadku, który wszystkich tutejszych obywatelów niezmiernym nabawił strachem. Wieczorem o 58. Minutach na 5. godzinę, dość znaczne trzęsienie ziemi tutejsze domy tak mocno wstrzęsło, iż na Ratuszu dzwon 6. albo 7. Razy zadzwonił; w jednym domu cynowe naczynia z miejsca poruszyły się; po innych domach krzesła z miejsc swoich poruszone zostały. W pewnym domu druciana klatka tak mocno wstrząśniona była, iż ptak w niej siedzący na drążku spadł na dół. Nawet o podal od Miasta, o tejże samej minucie uważano, iż most mocno wstrząsł się. Wielu tutejszych obywatelów twierdzi, że o tymże samym czasie, słyszeli wielki hałas i huk podziemny. Z Wyższego Szląska d.8. Grud: Dnia 3. tego miesiąca o 5. Wieczornej godzinie, tak tu, jako też w wielu okolicach, uczuć się dało znaczne trzęsienie ziemi, które na jednych miejscach mocniejsze, a na drugich lżejsze uważane było. Te trzęsienia ziemi szerzyły się aż do Glogau. W Neise dwa razy dzwony na Ratuszu zadzwoniły. W Rattibor domy rys dostały. A w Glogau trębacz na wieży uważał, iż ona się chwiała. Oprócz tego, żadnej szkody niemieliśmy.

LWOWSKIE TYGODNIOWE WIADOMOŚCI, Lwów:

No. 94, December 20, 1786

Z Warszawy dn. 9. Grudnia: z Krakowa donoszą że tam w Niedzielę przeszłą o godzinie piątej, i pięciu minutach po południu, w całym mieście, i jego okolicach dało się czuć trzęsienie ziemi, przez kilka Sekund trwające bez żadnej jednak ruiny znacznej prócz niektórych niewielkich rys. wysokość Barometru tego dnia w wieczór byłą 27 calów, Termometr stał w punkcie lodu. Czas tak przed trzęsieniem, jak i po trzęsieniu był cichy i pochmurny. Wiadomości odebrane z Pińczowa, Wawrzeńczyc, i z Nadzowa i z wielu innych miejsc Województw a Krakowskiego, i Sendomirskie zapewniają, że prawie w tym czasie, obydwa te Województwa czuły skutki tego trzęsienia ziemi. Z Polonca w Cyrkule Myślenickim piszą, że tam trzęsienie ziemi dn. 3go Grudnia o godzinie piątej, minutach 15. wieczór przypadło. Przy uspokojonym, i cichym bardzo powietrzu, lecz mglisty dał się słyszeć grzmot ziemny, na podobieństwo ciężkiego pojazdu po Grudnej i twardej ziemi spieszno idącego. Zaczęła się w tym momencie trząść ziemia, pierwej gwałtownie, ale nie długo, potym wolne ruszanie się ziemi dobrą minutę trwało. We wszystkich powszechnie domach pobliskich obserwowano toż samo, i widziano, że drzwi skołatane były, ściany i dachy trzeszczały, ale też bez wielkiej szkody. Dokładnie rozeznać można było, że to trzęsienie działo się od zachodu ku wschodowi.

No. 95, December 23, 1786

Trzęsienie ziemi dnia 3go Grudnia o pięciu minutach po godzinie piątej wieczór w Krakowie i w okolicach miasta tego przytrafione daleko mocniejsze, i straszniejsze nierównie było, niżeli to, które się tego roku dnia 27go Lutego stało. Pod ten czas Wisła straszliwie się wzburzyła, i na łokieć podniosła się. Dało się bardzo słyszeć, i czuć w Bochni, Sendomirzu, w Olkuszu, w Krzeszowicach, we wsi Wielkiejnocy 3 mil od Krakowa, Garnku o mil 3 od Częstochowy nad rzeką

Wartą leżącym, w Włoszczowie o mil 12 od Krakowa, i w Tyńcu. Oprócz jednak wielkiego kołatania rzeczy wszystkich po kamienicach znajdujących się, samych Kamienic, i budynków, i małych rys onychże o innej szkodzie nic nie słychać.

No. 96, December 27, 1786

Z Warszawy dn. 16. Grudnia: Dalsze wiadomości o trzęsieniu ziemi z listów późniejszych Krakowskich wybrane donoszą: że w Mieście Krakowie na wielu miejscach mury i sklepienia porysowały się, że po wielu miejscach około Krakowa pospadały z kościołów gzymsy, i Krzyże, że toż trzęsienie w Busku, w Kielcach, w Chęcinach, w Głębowicach, a czym bliżej Krakowa, tym mocniejsze było, że dalej, jak do Małogoszczy nie doszło; zatym albowiem miejscem, dążąc ku Warszawie czuć się nie dało.

No. 98, January 03, 1787

Z Warszawy dn. 23 Grudnia. Oprócz dawniej wyrażonych miejsc, ostatnim trzęsieniem ziemi kołatanym, piszą, że także w Kurozwękach w Województwie Sendomirskim, w Racimowicach, w Racławicach, w Zemboczynie, w Kowali, znacznie się to trzęsienie we znaki dało.

NOVA ACTA ACADEMIAE SCIENTARIUM IMPERIALIS PETROPOLITANAE, 1789, Petersburg, vol. V, 35–38

Extrait d'une Lettre de M. de Carosi, Capitaine au service du Roi & de la république de Pologne, & Correspondant de l'Académie, á Mogila près de Cracovie à Cracovie le 26 Janvier 1787

L'été passe ne m'ayant fourni rien qui soit digne de l'attention de l'Illustre Academie, je prends la liberté de communiquer un récit fidèle & circonstanciè du tremblement de terre que nous eumes en dernier lieu.

Après un été assés court, inconstant, per chaud & fort humide, & un automne qui lui repondoit, l'hyver dévança de beaucoup son temps ordinaire. Ce fut déja avant la mi – Octobre, que les gelées de nuit & les brumes nous surprirent: les premieres neiges tombèrent le 17 Octobre. Depuis ce temps les froids & les neiges continuèrent, en augmentant, & au commencement du mois de Novembre nous eumes des vents violens, approchans des ouragans, accompagnés d'une neige si copieuse, que tous les chemins en furent rendus impracticables pour plus d'une semaine, & qui recouvrit les habitations de nos villageois jusqu'à la moitié de leur hauteur. Quinze à 20 jours après survint un dégel rapide, qui fit déborder toutes les rivières & fleuves, déja bien couverts de grosses glaces : qui brisa, emporta les glaçons & rendit les rivières navigables. Le 3 Décembre à 5 heures 8 minutes du soir, il se fit entendre un bruit soûterrein bien fort, qui dura 5 secondes, & qui fut suivi immédiatement d'un tremblement de terre, dont la direction étoit du Midi au Nord, le Thermometre étoit précisement sur zero, point de la congelation de l'eau.

Ce ne furent point de secousses marquées, mais plutôt des balancemens bien forts, vers les deux regions marquées, qui durèrent 10 secondes à peu près. Ce fut le plus fort tremblement qu'il y ait eû dans ces contrée depuis une temps immémmorial, néanmoins il n'a occasioné aucun dommage remarquable. Des tasses, des verres &c. se remuèrent sur les tables, commodes & armoires; des tableaux, des cages d'oiseaux attachés aux parois, balancerent ça & là & il y en eut même de tombés par terre; des portes ouverte se fermerent, d'entre-ouvertes s'ourvrirent, & celles qui étoient bien fermées craquerent & furent ébranlées. Il y eut aussi quelques morceaux de corniche à demi détachés de jettés à terre. Bien de personnes observèrent très - distinctement l'inclinaison & le redressement des parois de leur chambres d'autres des chéminées & toits. Là où le tremblement se fit sentir plus fort qu'ailleurs, (c. a. d. où les couches pierrenses sont plus près de la surface,) les personnes surprises debout, eurent de la peine à se tenir sur pieds, & il y en eurent même de renversées. Comme la Vistule étoit redevenue, navigable par le dégel, il y avoit des vaisseaux chargés de marchandises dépêchès de Cracovie pour Varsovie, & au moment du tremblement les bâteliers virent le fleuve se gonfler & sentirent un chancellement de leurs navires très marqué. Par le débordement des rivières & ruisseaux & par les gelées survenues aprés, les plaines atténantes avoient été couvertes d'une croute de glace, qui au moment du tremblement se brisa avec éclat; desorte, que non seulement les passans en furent épouvantés, mais que ceux – mêmes qui l'entendirent à une certaine distance, en furent troublés.

Pour ce qui est des païs attenans, comme la Gallicie, voice ce que j'en ai pu apprendre : le tremblement de terre fut sent en beaucoup d'endroits avec une force inégale, à Wieliecka surtout toute la ville étoit en crainte, & dans les soûterreins des mines à sel il y eut un gros bloc de détaché & écroulé, mais les maisons n'en furent point endommagées. Le meme bruit soûterrein qui précédoit le tremblement, & puis les secousses vacillantes, se firent sentir à Tarnow, Myslenie, Zator, Oswiecim, Bilie, Biala & en bien d'autres endroits; pour ce qui est des Carpates proprement dites, il m'a été impossible d'en tirer quelques informations, quoique il doit y avoir absolument eû des secousses, vû qu'il y a plusieurs volcans éteints, comme Babia Gora & bien d'autres. Qu'on l'ait senti dans la haute Silesie, tant autrichienne que prussienne, les gazettes nous l'ont appris, de même que de la Hongrie, ce qui donne d'autant plus lieu de croire que les Carpates n'en ayent point été examptes. L'étendue en long, c. a. d. du Midi au Nord, où cet évènement fut senti chez nous, fait à peu près 14 à 15 lieues d'Allemagne, savoir depuis Cracovie jusqu'à Piotskow, derriere Malogoszer, Kielce, Pinerow, & lea àlentours. Au delà dans les montagnes & mines à fer, comme à Radoscyce, Konskie, Drzewica, &c. on n'en a plus rien senti. L'étendue en large est à peu près de même que la longueur.

Voilà en 14 mois le troisieme tremblement de terre chez nous, évènement qui depuis des siecles fut inconnu è nos ancêtres, n'étant arrivé que par de long intervalles. Il seroit curieux de connoitre l'origine de ce phénomène: j'ai attentivement lu les gazettes, mais je n'y trouve que le seul 24 Novembre de l'année passee. Où il y eut un tremblement de terre à Rome; le nôtre en seroit – il une suite, ou auroit il eu une autre source ?

Les suites de ce tremblement de terre furret chez nous : que le temps changea presque aussitôt : nous eumes encore des plus forts dégels, des neiges, de la pluye & des brouillards extraordinaires, dont le plus fort de tous fut remarqué le 10 Décembre, entre 2 & 3 heure après midi; il monta à l'Est & remplit tellement l'atmosphère, qu'elle en fut obscurcie comme après le coucher du Soleil; il étoit aussi épais qu'une grosse fumée: desorte qu'à Varsovie il y eut bien de personnes qui crurent qu'il y avoit du feu en ville. Les brouillards precédens & suivans furent tous bien moindres, ils se léverent tous après midi à la même heure & ne durerent qu'à peu près 1 1/2 ou 1 3 /4 d'heure. Ce temps nébuleux & humide ayant continué 13 jours, il fut suivi par des gelées & un ciel serein, qui continua presque sans interruption jusqu'à la fin du mois.

SCHLESISCHE PROVINZIALBLÄTTER, 1786, Oleśnica, vol. 4, No. 12, 593–596
Nachricht von dem Erdbeben am 3. Dezember d. J.
In einem Zeitraum von ungefähr 16 Monaten, hat unser Schlesien nun schon zu drei wiederholten malen diese große Naturbegebenheit erfahren und ihre Würkung scheint sowohl an Heftigkeit als an weitere Umfange, bey jedesmaliger Rückkehr nur noch mehr zuzunehmen. So schreckhaft aber auch überhaupt genommen diese Würkung der Natur für unsere Landesleute ist, so ist sie aber doch nicht so ganz ungewöhnlich, als der größte Theil derselben davor hält. Aus physischen Ursachen können auch Erderschütterungen in einem Lande nicht so ganz ungewöhnliche Erscheinungen seyn, das so unleugbare Beweise ehemaliger Vulkane in seinen Grenzen aufbewahrt, wie unser Schlesien. Unsere Chroniken enthalten verschiedene Nachtrichten von Erderschütterungen, die in den vorigen Jahrhunderten in unseren Gegenden bemerkt wurden oft war sie noch heftiger als diese neueren, aber doch zum Troste unser bekümmerten Mitbürger nie so stark, dass Häusern oder ganze Städte dadurch in Ruinen verwandelt worden wären. Auf nähere Angaben kann ich mich hier nicht einlassen, da der enge Raum mir nur erlaubt, die vorzüglichsten Begebenheiten dieser neuen Erschütterung am verflossen 3. Dec. d. Jahres ganz kurz zu berühren. Der Würkungskreis dieser wiederholten Erderschütterung erstreckte sich nicht nur auf eben die Gegenden, wo sie am 27. Februar verspürt wurde, sondern auch auf einen größeren Theil von Niederschlesien und sogar der angrenzenden Orte des Nordischen Pohlens. Die Richtung der Stöße war so wie damals, und seine größte Stärke ward ebenfalls um Tarnowitz, Pless, Ratibor, Cracau und andern Gegenden des südlichen Pohlens bemerkt: (man sehe Provinzialblatt d. J. März, S. 202 u. s. w.) doch scheint es diesmal die nördlichen Gegenden von Böhmen, Mähren und Ungarn nicht betroffen zu haben, wenigsten seien nähere Nachrichten aus diesen Gegenden gänzlich; desto weiter aber verbreitete es sich über die nördlichen Grenzen Schlesiens.
Alle Nachtrichten stimmen darin überein, dass die Erschütterungen diesmal weit heftiger als im Februar waren. Und da sie sich nach Verschiedenheiten der Uhren jedes Ortes kurz vor oder bald nach 5 Uhr Nachmittags einstellten, so konnten sie auch von einem sehr großen Theil der Einwohner dreier Gegenden bemerkt werden. Der Himmel war zu dieser Zeit so wie den ganzen Tag über trübe, in der Luft aber herrschte eine ungewöhnliche Stille, den Tag über wehte der Wind meißt aus N. O. Kurz vorher oder vielmehr während den drei Stößen die in 6 bis 8 Sekunden aufeinander folgten, hörte man an den mehresten Orten ein sehr

starkes Geräusch oder Brausen in der Luft, womit das Klirren der Fenster, Gläser und Schüttern der Thuren und Häuser begleitet war, und während die 3 Stöße in kurzen Zwischenräumen auf einender folgten, davon der zweite der heftigste war. In Tarnowitz verließen viele Einwohner ihre Wohnungen und eilten auf den Markt oder ins Freye. Viele Häuser daselbst so wie auch in Plesse, und einigen Orten in Pohlen, (davon die öffentlichen Nachrichten umständliche Berichte enthalten) bekamen durch diese heftige Erschütterungen starke Risse in den Mauern, die von oben bis in den Grund gehen. In einem Hause stürzte die Kappe vom Schornsteine in denselben hinein, und zerschlug eine Menge Küchengeräte. An mehreren anderen Häusern wurden die Gesimse der Schornsteine zum Theil herabgerissen, oder doch sehr stark beschädigt. Das Prassel der Balken, Klirren der Fenster, Schüttern der Thüren, und herabfallen des Schuttes und Ziegelstücken, machte das Schrecken und diese fürchterliche Naturszene unter den Bewohnern dieses und aller umliegenden Orte nur noch allgemeiner, da man es in Beuthen, Neudek und anderen Orten noch häufiger als in Tarnowitz selbst verspürte, auch in Plesse war die Wirkung fast eben so heftig, viele Häuser bekamen auch da Risse, und vorzüglich wurden die Stubenofen dergestalt zerrüttet, dass sie des andern Morgens des eindringenden Rauches wegen, nicht geheizt werden konnten. Aus Ratibor, Neustadt, Neisse, Krappitz, Cosel, Ober-Glogau, Oppeln, Brieg, Loewen, Stolz, Münsterberg und anderen Orten, war die Erschütterung zwar auch stark, doch nicht so heftig als an den vorigen Orten. In Carlsruhe schlug der Hammer an der Uhrglocke sowohl auf dem Schloss, als der Kirche verschiedene male gedämpfte Stimme, und der Adler auf der Turmspitze drehte sich beständig während dieser Erschütterung. Das Anschlagen des Hammers an die Glocken, ward auch in Namslau und einigen anderen Orten bemerkt. In Creuzburg, Rosenberg, Pitschen, Landsberg und Constadt wurden die Stöße gleichfalls verspürt doch in noch geringeren Graden als in vorherbeachenden Orten. Eben so beachtete man die Erschütterung in Oels, sowohl auf dem Herzogl. Schlosse als auch an einigen Orten der Stadt. In Bernstadt waren sie schon stärker als hier, weit schwächer aber in den Gegenden Juliusburg, Trebnitz, Festenberg, Militsch, Sulau, Zduno, Rawitsch und anderen Orten mehr. An den mehresten dieser Orte bestand die Wirkung dieser Erschütterung außer dem Schrecken derer, die sie bemerkten, aus einem leichten Schüttern der Stühle, Tische, Fenster und Stubenthüren und allen freistehenden Sachen. Bey Vögeln und anderen Thieren wurden auch diesmal an vielen Orten, so wie im Februar gleiche Empfindungen der Furcht bemerkt. In Breslau empfanden diese Erschütterung auf dem Dom, der Schubrücke, im Ohlauischen Viertel, auf der Albrechtgasse, auf dem Naschmarkte, auch verschiedene alleinstehende Personen. Die mehresten bemerkten nur eine sehr leichte Erschütterung der Thüren und Erklingen der Tassen und Gläser auch in Liegnitz und Glogau will man sie bemerkt haben. Bald nach dem diese Erschütterung vorüber war, fing es an den bey noch fortdauernder stillen Luft etwas zu heftiger Nebel, der bis 9 Uhr dauerte, und zu Anfange bis nach 6 Uhr mehrentheils auf der Oberfläche der Erde zu ruhen schien. Indem er dichter wurde, je naher er der Erde war, und durchsichtiger, jemehr er sich davon entfernte dass auch der damals sehr hell hindurch scheinende Mond ein sehr sonderbares Schauspiel dem Auge darbot. Gegen 9 Uhr wollen einige eine widerholte schwache

Erschütterung bemerkt haben, und gegen halb 10 Uhr stellte sich etwas heftiger Sturm mit einigen starken Stößen ein. Die würkende Ursache dieser Erschütterung, wollten einige Freunde der Natur zwar einer fernen großen Revolution der Erde zuschreiben, davon uns aber erst nähere Nachrichten mit Gewissheit belehren können. Vielleicht dürften jedoch diese Vermutungen, so wie bey dem Erdbeben im Februar bloße Vermutungen bleiben. Da aber diese Erschütterungen ihre größte Kraft um Tarnowitz, Pless und vorzüglich Beuthen geäußert, so glauben andere, sie vielleicht mit mehrerem Recht in diesen Gegenden selbst suchen zu müssen. Der letzte Ort ist seines ehemaligen starken Bergbaues wegen berühmt; Könnten sich nicht, dachten sie, in diesen alten meisst verfallen Schachten, unterirdische Dünste in solcher Menge gesammelt haben, dass sie durch diese Gewaltsame Explosion ihre Zertheilung gesucht hatten? Ob zwar überhaupt genommen, eine gewaltsame Zertheilung der Dünste auf diese Art nichts unwahrscheinliches gegen sich hat, da sie tägliche Erscheinung ist, so hatte, im Fall sie wirklich stattgehabt, der Wirkungskreis dieser Erschütterung doch von einem so ausgebreiteten Umfange seyn können. Er würde sich bloß auf einen kleinen Bezirk der umliegenden Gegend verbreitet, und seine Stärke auf eine vielleicht noch verwüstendere Art gezeigt haben, da seine würkende Kraft, in einer geringen Tiefe, und gleichsam nur auf der Oberfläche der Erde lag. Es scheint mir daher wahrscheinlicher, dass so wie ich auch damals im Februar annahm, die würkende Kraft dieser Erschütterung, zwar hauptsächlich in diesen Gegenden, aber viel tiefer in dem Innern der Erden befindlich gewesen sein müsste. Sollten sie nicht viel mehr die wirklichen, ob zwar nur noch schwachen Überreste der Würkungskraft unser ehemaligen Vulkane seyn? Die ebenso sonderbaren als außerordentlichen Naturbegebenheiten, die sich seit einigen Jahren hauptsächlich in unseren Gebirgsgegenden zutrugen, wovon uns der Naturkalender für Schlesien, und die Phys. Zeitungen verschiedene aufbehalten haben, scheinen mir wahre Tatbeweise dieser Zumuthung zu seyn.

Oels, gedruckt bey Samuel Gottlieb Ludwig, Herzogl. Hofdrucker.

SCHLESISCHE PRIVILEGIRTE ZEITUNG

No. 145, 9 December 1786

Breslau den 7. Dec. Verschiedene Personen hatten am Sonntag Nachmittag gegen 5 Uhr eine Erdschütterung bemerkt, die man doch nach näherer Untersuchung erst prüfen wollen; allein die von anderen Orten in Schlesien eingegangen Berichte setzen diese große Naturbegebenheit außer Zweifel. Hier in Breslau ist sie besonders von der Seite nach der Ohlauischen Vorstadt, dem Dohm und Sand bemerkt worden. Einige glaubwürdige Personen versichern sogar Bewegungen der hängenden Uhren und stehenden Gläsern bemerkt zu haben. In Brieg, Neisse, Leobschütz ist es stärker gewesen. In Ratibor ist eine 2 Fuß dicke Mauer gesprungen und in verschiedenen Häusern sind Gläser heruntergefallen. Man wird die nähern Nachrichten nicht ermangeln mitzutheilen.

Namslau den 4. Dec. Unsere Stadt hat am gestern Abend eine Naturbegebenheit erfahren, welche alle hiesige Einwohner in Schrecken versetzte. Es war Abends,

56 Minuten auf 5 Uhr, als ein ziemlich heftiger Erdstoß die hiesigen Gebäude, vorzüglich in dem gegen Morgen liegenden Theile der Stadt, dergestalt erschütterte, dass der Kleppel aus der Schlagglocke des Ratsthurms 6 bis 7 mal anschlug, das in einem gewissen Zimmer auf einem Rechen befindliche zinnerne Gefäße von seiner Stelle gerückt wurden, die nicht verschlossenen Schrankenthüren in vielen Häusern aufsprungen und bewegten, die Stühle mit denen so darauf saßen, hin und her wankten, und ein Vogelbauer von Drath in einem andern Zimmer so stark bewegt wurde, dass der darinnen befindliche Vogel von dem Stängel, worauf er stand, herunter fiel. Auch weit von der Stadt bemerkte man auf einer Brücke, in eben derselben Minute eine heftige schwankende Bewegung, und viele hiesige Einwohner wollen um diese Zeit ein starkes Getöse unter der Erde wahrgenommen haben. Wahrscheinlich muss in einer entfernten Gegend wieder eine heftige Erdschütterung vorgefallen sein, und es ist zu wünschen, dass nicht traurigere Nachrichten auf diese unsere bange Bemerkung folgen.

Aus einem Schreiben aus den Rattiborschen, den 3. Decemb. um 6 Uhr Abends. Heute um 5 Uhr 10 Minuten Nachmittag, haben wir hier wieder ein Erdbeben verspürt. Man bemerkte nichts als ein Geräusch in der Luft und gleich darauf einen Stoß, welcher ungefähr 4 Sekunden gedauert hat, die Richtung war wieder von Osten nach Westen. Die das vorherige, welches im Monat Februar war, verspürt haben, behaupten er sey nicht so stark gewesen, doch war es so, dass Tintenfass vom Tische herunter gefallen wäre, wenn solches nicht noch aufgehalten worden. Bücher so auf dem Tisch auf einem Haufen gelegen, sind herunter gefallen, wie auch Zeitungen. Der Barometer stand den ganzen Tag auf Veränderlich, auch ist er nach dem Stoß, wo ich gleich darauf gesehen hatte, eben so geblieben, nur jetzt scheint er etwas weniges zu steigen, auch wird es etwas kälter. Der Himmel war den ganzen tag gewölkt. In der Stadt Riebnick haben sie es eben bemerkt. Die Schildwacht versichert, der Rathsthurm habe gewankt, auch verschiedene Bürger sagen, dass die Dächer geknastert hätten, indessen ist auf der Leute ihre Aussage nicht viel zu trauen, da die Angst bey dergleichen Leuten viel macht. Auch ist noch zu bemerken, dass, da das Schloss hier mit Wasser umgeben, das Eis verschiedene Knalle gethan hat.

No. 146, 11 December 1786

Auszug eines Schreibens aus Plesse, den 6. Dec. Sonntags den 3ten dieses, Abends, um Viertel auf 5 Uhr verspürten wir hier eine ziemlich starke Erderschütterung, dass darüber die meisten Leute aus Häusern heraus gelaufen; denn wirklich, auch einige Häuser bekamen Risse und die Stuben-Öfen wurden dergestalt zerrüttet, dass den andern Morgen wegen des eindringenden Rauche die Zimmer nicht geheizt werden konnten. Hühner und Vögel sind von ihren Sitzen herunter geworfen, und an manchen Orten auch die Gläser. Die Erschütterung schien nicht so stark als im Februar dieses Jahres gewesen zu seyn, indessen spürte man doch die Würkung stärker. Leute so sich eben zu dieser Zeit auf dem Felde befunden, versichern übereinstimmend, dass es gewesen wäre, als wenn sie mit der Erde hätten herunter sinken sollen und wieder hernach in die

Höhe geworfen wurden. Auch aus Cracau ist die Nachricht eingegangen, dass das Erdbeben um 5 ¼ Uhr Sonntags Abends ebenfalls gewesen, und 2 Sekunden gedauert haben soll, jedoch aber Gottlob! Keinen Schaden verursachet. Nach dem Zeugniss des Thurmwächters ist es aber doch auch stärker als sonsten gewesen, indem der Thurm so stark bewegt worden, dass der Seeger zweimal angeschlagen. Aus dem benachbarten Kayserlichen haben wir noch weiter nichts gehört, als dass das Schloss in Drahemyschel dermaßen geschüttelt worden, dass davon Stücken von der Mauer herunter gestürzt, wodurch ein Mensch stark beschädigt worden.

Auszug eines Schreiben aus Tarnowitz, vom 4 December. Gestern, als Sonntags gegen Abend, als ich eben bey Herrn K. war, hatten wir hier ein Erdbeben. Es war nach 5 Uhr Abends, und ich saß mit ihm allein auf seiner Stube, nach ¼ auf 6 Uhr fing der Stuhl auf dem ich saß sich unter mir so zu bewegen an, dass es nicht allein empfunden, sondern auch gesehen werden konnte. Ich fragte den mir gegen über sitzende Freund: Fühlen sie nichts? Er sprang auf, und sagte: Lassen uns ins Freye gehen; indem fingen Die Wände an zu wanken, die Fenster zu klirren, Decke über uns und Boden unter uns sich sichtbar zu bewegen; eine Menge Staub erfüllte das Zimmer, und wir eilten der Thüre zu, die ich aber nicht bald aufmachen konnte, weil sie noch durch die fortwährende Bewegung geklemmt ward. Das war unser Glück, denn während dem fiel ein Ziegel und mehr Schutt herunter, gerade auf die Treppe, wo wir ungefähr gewesen wären, wenn die Thür gleich aufging. Wir eilten ins Freye zu kommen, und fanden auf dem Markte schon mehrere die ihre Wohnungen verlassen hatten; ein jeder erzählte was er erfahren, gefühlt, gesehen hatte. Das Wetter war sehr stille und trübe; aber gar nicht kalt. Der Wind stand aus Nordost. Die Stöße kamen aus Südwest, folglich dem Winde gerade entgegen in dem nehmlichen Richtung, wie den 27. Febr. aber ungleich stärker, denn es ist fast kein Haus am Markte, besonders, die wir gegen Morgen, Mittag und Abend liegen, die nicht etwas gelitten hätten. In einem Hause stürzte die Kappe vom Schornsteine in denselben hinein und zerschlug eine Menge Küchengeräte. An mehreren Häusern sind die Gesimse von Schornsteinen abgerissen. Von auswärts haben wir noch wenig Nachricht, nur so viel, dass es in Lubschau, Neudeck und in allen uns umgebenden Dörfern gespürt ward. Das Schicksal der umliegenden Städtchen wissen wir noch nicht; und ich wünsche dass nirgends wo mehr Schaden geschehen seyn möchte als bey uns. Auf unsern Gruben ist es auch verspürt worden, allein da es eben Sonntag war und niemand in den Gruben arbeitete, so weiß man von der Bewegung unter der Erde nichts. Auf der Schmelzhütte haben die Arbeitenden nichts gemerkt, aber in den nahe dabey liegenden Rybna war die Erschütterung noch heftiger als in der Stadt. Die Zeit des Erdbebens war 8 bis 10 Secunden, während dieser Zeit kamen 3 Stöße.

No. 150, 20 December 1786

Warschau den 13. Dec. Die Nachrichten vom Erdbeben in diesem Königreiche sind sehr übereinstimmend in Absicht der Zeit, und in diesem Jahre sind bereits zwey bemerkt worden, ohnerachtet es nichts seltenes ist, nur oft ist der

Zwischenraum lang und kurz gewesen. Pohlnischen Geschichtsschreiber setzen das erste Erdbeben Anno 1000, das 2te An. 1016, das 3te Anno 1200, das 4te An. 1257, das 5te An. 1303, das 6te An. 1348, das 7te An. 1358, das 8te An. 1785, das 9te 1786 den 27 Febr. und das 10te 1786 den 3ten Dec.

Ein Schreiben aus dem Cracauischen sagt folgendes. Den 3ten Decemb. 5 Minuten nach 5 Uhr Abends hörte man ein unterirdisches Getöse, als wenn Wagen auf der Straße führen. Zu Kazimierz so nach Mitternacht liegt, wurden alle Häuser bewegt, die Dächer knasterten, die Stühle und alles, was in dem Zimmer war, bewegte sich, die Thüren der Stuben und Keller sprangen auf, die Uhren schlugen an, die Soldaten in der Wachstube retirierten sich, weil solche einzufallen drohete. In der Kirche zur heil. Maria war alles in Bewegung, die Geistlichen, so im Refectorio befindlich, entfernten sich, während dieser Zeit bewegte sich die Weichsel mit außerordentlicher Gewalt, Leute die über die Brücke gingen, glaubten mit samt der Brücke unterzugehen. In Bochnia sind Ofen eingefallen, und das Feuer in den Stuben herumgeworfen worden. In vielen Zimmern sind die Balken aus einander gegangen, die Uhr hat unzähliche mal angeschlagen. In Sendomir ist bey den Reformaten das Kloster gesprungen, Gläser und alles was auf dem Tisch gelegen, herunter gefallen, und auch hier hat die Weichsel außerordentliche Bewegungen gemacht. Zu Olkuz sind die Leute mit dem Stuhle umgefallen. In Krzesowicza hat die Kirche Schaden gelitten, und die Glocken haben sehr heftig angeschlagen. In dem Dorfe Wielkieynoc, 5 Meilen von Cracau sind in dem Schlosse Wandleuchter und Spiegel herunter gefallen, und ist alles in Bewegung gewesen. In Garnka 5 Meilen von Czenstochau an der Warte gelegen, sind die Wände geborsten. In Wloszowa 12 Meilen von Cracau ist das ganze Städtchen in Furcht gerathen, weil alles sich bewegt, Thüren und Fenster aufgesprungen, die Kirche hat dabey das meiste gelitten, auch in Tyniec ist es sehr heftig gewesen. Zu Opatow hat man es auch gemerket, besonders haben die Gegenden des Weichselstroms es am heftigsten empfunden. Zu Cracau sind ein paar Vasen von einem Hause gefallen.

Bielitz den 13 Dec. Den 3. Dieses Abend um 5 Uhr hatten wir eine mit starkem Brausen begleitete ziemlich heftige Erderschütterung. Binnen 14 Monaten ist das die dritte, und eine in unsern Gegenden sonst nicht bekannte Begebenheit. Sie hat zur nehmliche Stunde weit nach Gallizien und Ungarn getroffen; im Carpathischen Gebirge stärker als hier.

No. 151, 23 December 1786

Warschau den 16 Dec. Einige Nachrichten vom Erdbeben sind noch nachzuholen. Im Carmeliter Kloster zu Cracau war ein Geistlicher welcher seit 3 Wochen auf keinen Fuss treten konnte, sondern beständig im Bette zubrachte, bey der ersten Erschütterung übermannte ihn die Furcht, so das selbiger mit einem Sprung aus dem Bette war und davon lief, und ist bis jetzt von allen Schmerzen befreyet. Die Augustinerkirche ist gesperret, weil solche so starke Risse bekommen dass sie einzufallen drohet. Es hat überhaupt hin und wieder wichtige Störungen in den Gemäuren gemacht.

5.3 Catalogue of Localities for Which the December 3, 1786, Earthquake Evidence Was Available. (Quotations Translated into English)

Barczków (Baczków), Poland

1. Uście Solne – manuscript
The earthquake was felt in Barczków as much as in Uście Solne. For detailed description of the event see: Uście Solne.

2. Bójko, 1911, p. 49
mention after Uście Solne - manuscript

Balice, Poland

1. Gazeta Warszawska, 1786, Dec. 16, No. 100
*…At **Balice,** one villager went to fetch water from a spring when a rock split in two parts with such a noise as though several guns were fired at once next to his ear…*
(Balice is now a suburb of Kraków)

2. Gazeta Wileńska, 1786, Dec. 23, No. 51
The same repeated after Gazeta Warszawska, 1786, Dec. 16, No. 100

Bielsko-Biała (Bielitz), Poland

1. Bielsko-Biała AP, manuscript
…On December 3 at 5 p.m. all around here the earthquake was felt again. It lasted a quarter of a minute, and, after that, fortunately, God be thanked for no misfortune happened…

2. Schlesische Privilegirte Zeitung, 1786, Dec. 20, No. 150
…On 3 December at 5 p.m. we experienced quite a strong tremor accompanied by strong noise…

3. Carosi, 1789, p. 37
*…The same underground noise - which preceded the earthquake - and the next shakes could be felt in Tarnów, Myślenice, Zator, Oświęcim, **Bielsko-Biała**, and many other places …*

Bieńkowice, Poland

1. Uście Solne - manuscript
The earthquake was felt at Bieńkowice as much as at Uście Solne. For detailed description of the event see: Uście Solne.

Bierutów (Bernstadt), Poland

1. Schlesische Provinzialblätter, 1786, p. 595
*…It was stronger in **Bierutów** (than in Oleśnica), however it was far weaker in area of Dobroszyce, Trzebnica, Twardogóra, Milicz, Sułów, Zduny, and Rawicz.*

In most of these localities - apart from the fear of those who experienced it - it was manifested by the slight shaking of chairs, tables, windows in rooms, and all standing objects. Also at this occasion, it was noticed that birds and other animals showed the same signs of fear as those in February...

Bochnia, Poland

1. Uście Solne - manuscript
... *It was started in Bochnia a half an hour before sunset...*

2. Gazeta Warszawska, 1786, Dec. 13, No. 99
...*There was such a great earthquake at **Bochnia** that through the tiles separating in furnaces one could see the fire; and as soon as the earthquake stopped the tiles came together again. In some rooms doors opened. At the Castle, and some other places, the clock bells rang a few times on their own...*

3. Gazeta Wileńska, 1786, Dec. 23, No. 51
The same repeated after Gazeta Warszawska, 1786, Dec. 13, No. 99

4. Lwowskie Tygodniowe Wiadomości, 1786, Dec. 23, No. 95
...*It was heard and felt very much at **Bochnia**, Sandomierz, Olkusz, Krzeszowice, Wielkanoc 3 miles from Kraków, Garnek on the Warta river 3 miles from Częstochowa, Włoszczowa 12 miles from Kraków, and Tyniec...* (1 Polish mile = 7146 m)

5. Schlesische Privilegirte Zeitung, 1786, Dec. 20, No. 150
following Gazeta Warszawska but much exaggerated
...*At Bochnia furnaces broke and the fire spread to rooms. In many rooms beams separated. A clock struck an infinite number of times...*

6. Brünner Zeitung, 1786, No. 104
after Schlesische Privilegirte Zeitung, exaggerated

7. Jeitteles, 1860, p. 329
the same, following Brünner Zeitung

Bogucice, Poland

1. Uście Solne - manuscript
The earthquake was felt at Bogucice as much as at Uście Solne. For detailed description of the event
see: Uście Solne.

2. Bójko, 1911, p. 49
mention after Uście Solne - manuscript

Bratucice, Poland

1. Uście Solne - manuscript
The earthquake was felt at Bratucice as much as at Uście Solne. For detailed description of the event
see: Uście Solne.

2. Bójko, 1911, p. 49
mention after Uście Solne - manuscript

Brudzów, Poland

1. Kraków AKMK-6, manuscript
*...at the **Brudzów** village the second time the earthquake was felt on December 9 at 8, at sunrise...hooks at doors moved, locks opened, hanging bars knocked, window-panes rattled...*

Brzeg (Brieg), Poland

1. Schlesische Privilegirte Zeitung, 1786, Dec. 9, No. 145
*...In **Brzeg**, Nysa and Głubczyce the earthquake was stronger...(than in Wrocław)*

2. Schlesische Provinzialblätter, 1786, p. 595
Brzeg is only mentioned that there the earthquake was strong but not as much as at Bytom, Tarnowskie Góry, Świerklaniec and Pszczyna.

3. Gazeta Warszawska, 1786, Dec. 30, No. 104
the same following Schlesische Privilegirte Zeitung

4. Hoff, 1841, p. 78
mention

5. Jeitteles 1860, p. 330
mention after Schlesische Privilegirte Zeitung

Busko, Poland

1. Gazeta Warszawska, 1786, Dec. 16, No. 100
*...There was also the earthquake in **Busko**, Kielce, Chęciny, Głębowice and the whole of the Zator and Oświęcim Duchies and closer to Kraków than stronger. In Siosławice, a few "staja" from Busko, the newly erected buildings showed so many cracks that their inhabitants had to leave their houses in a hurry....* (a Polish staja = ca 130 m)

2. Gazeta Wileńska, 1786, Dec. 23, No. 51
The same repeated after Gazeta Warszawska, 1786, Dec. 16, No. 100

3. Lwowskie Tygodniowe Wiadomości, 1786, Dec. 27, No. 96
after Gazeta Warszawska

Byczyna (Pitschen), Poland

1. Schlesische Provinzialblätter, 1786, p. 595
Byczyna is only mentioned that there tremors were felt but to a lesser extent than in Brzeg, Nysa, Koźle, Racibórz, Prudnik, Krapkowice, Głogówek, Opole, Lewin Brzeski, Stolec and Ziębice.

Bytom (Beuthen), Poland

1. Schlesische Provinzialblätter, 1786, p. 594

...In Bytom and Świerklaniec the earthquake was stronger than in Tarnowskie Góry itself ... The earthquake was experienced the strongest in Tarnowskie Góry, Pszczyna, but especially in ***Bytom****...*
see: Tarnowskie Góry

Cerekiew, Poland

1. Uście Solne - manuscript
The earthquake was felt at Cerekiew as much as at Uście Solne. For detailed description of the event
see: Uście Solne.

2. Bójko, 1911, p. 49

mention after Uście Solne - manuscript

Chęciny, Poland

1. Gazeta Warszawska, 1786, Dec. 16, No. 100
Chęciny is only mentioned that there the earthquake was felt.

2. Gazeta Wileńska, 1786, Dec. 23, No. 51
The same repeated after Gazeta Warszawska, 1786, Dec. 16, No. 100

3. Lwowskie Tygodniowe Wiadomości, 1786, Dec. 27, No. 96
the same after Gazeta Warszawska, 1786, Dec. 16, No. 100

Cieszyn (Teschen), Poland

1. Cieszyn AP, manuscript:
...It came back with even greater force on 27 February ...on the first Advent Sunday, 1786 (i.e., on 3 December) *at about 5 p.m. The two earthquakes were so violent that ...as a result of the second one, boxes with files collapsed in the county office, and a piece of the town walls at the Higher Gate fell out...*

2. Jeitteles, 1860, p. 330
...In ***Cieszyn*** *the earthquake was felt to be stronger than this of February 27 of the same year and that of August 22, 1785 ... A piece of the wall broke away from the Higher Gate by itself...*

3. Gwiazdka Cieszyńska, 1929, No. 28
... On the third Monday of December at 5 p.m. the town and the environs were frightened by the third tremor. The strongest shock came at the Higher Gate, where a piece of the town wall broke away and the house of Baron Skribenski was greatly damaged...

Dobroszyce (Juliusburg), Poland

1. Schlesische Provinzialblätter, 1786, p. 595
...It was stronger in Bierutów (than in Oleśnica); however, it was far weaker in area of ***Dobroszyce****, Trzebnica, Twardogóra, Milicz, Sułów, Zduny and Rawicz. In*

most of these localities – apart from fear of those who experienced it – it was manifested as the slight shaking of chairs, tables, windows in rooms and all standing objects. Also at this occasion, it was noticed that birds and other animals showed the same signs of fear as those in February...

Drogomyśl (Drahemyschel), Poland

1. Schlesische Privilegirte Zeitung, 1786, Dec. 11, No. 146
*...The castle at **Drogomyśl** shook so much that pieces of walls fell out, badly injuring one man...*

Drzewica, Poland

1. Carosi, 1789, p. 37
*...Further away in the mountains and iron mines, like Radoszyce, Końskie, **Drzewica*** etc., *nothing was felt...*

Garnek, Poland

1. Gazeta Warszawska, 1786, Dec. 13, No. 99
*...In **Garnek** on the Warta river 3 miles from Częstochowa chests and stools in rooms were shaking, the people on the pavement were raised and those leaning against walls were pushed outwards...*

2. Gazeta Wileńska, 1786, Dec. 23, No. 51
The same repeated after Gazeta Warszawska, 1786, Dec. 13, No. 99

3. Lwowskie Tygodniowe Wiadomości, 1786, , Dec. 23, No. 95
Garnek is only mentioned that there the earthquake was felt and heard very much.

4. Schlesische Privilegirte Zeitung, 1786, Dec. 20, No. 150
*...In **Garnek** on the Warta river, 3 miles from Częstochowa, walls broke...*
information probably after Gaz.War., greatly exaggerated, 1 Polish mile = 7146 m

5. Brünner Zeitung, 1786, No. 104
the same after Schlesische Privilegirte Zeitung, 1786, Dec. 20, No. 150

6. Jeitteles, 1860, p. 329
mention after Brünner Zeitung

Głębowice, Poland

1. Gazeta Warszawska, 1786, Dec. 16, No. 100
Głębowice is only mentioned that there the earthquake was felt.

2. Gazeta Wileńska, 1786, Dec. 23, No. 51
The same repeated after Gazeta Warszawska, 1786, Dec. 16, No. 100

3. Lwowskie Tygodniowe Wiadomości, 1786, Dec. 27, No. 96
the same after Gazeta Warszawska, Dec. 16, No. 100

Głogów (Glogau), Poland

1. Gazeta Warszawska, 1786, Dec. 30, No. 104
...These earthquakes reached to ***Glogau*** *...and there, in* ***Glogau****, the bugler in the tower thought that it shook...*

2. Schlesische Provinzialblätter, 1786, p. 595
...In Wrocław most people only noticed the very slight shaking of doors and that cups and glasses rattled. The same was noticed in Legnica and ***Głogów****...*

Głogówek (Ober-Glogau), Poland

1. Schlesische Provinzialblätter, 1786, p. 595
Głogówek is only mentioned, giving information that the earthquake was strong but not as much as in Bytom, Tarnowskie Góry, Świerklaniec and Pszczyna.

Głubczyce (Leobschütz), Poland

1. Schlesische Privilegirte Zeitung, 1786, Dec. 9, No. 145
...In Brzeg, Nysa and ***Głubczyce*** *the earthquake was stronger... (than in Wrocław)*

2. Gazeta Warszawska, 1786, Dec. 30, No. 104
the same after Schlesische Privilegirte Zeitung

3. Hoff, 1841, p. 78.
mentioned

4. Jeitteles, 1860, p. 330
mention after Schlesische Privilegirte Zeitung, 1786, Dec. 9, No. 145

Gnojno, Poland

1. Kraków AKMK-6, manuscript
...They say that in the whole ***Gnojno*** *parish there was a great earthquake, that children fell from stove-corners, and that the stoves cracked, and the tops of buildings creaked. And even poultry fell from their perches. At Widuchowa of* ***Gnojno*** *parish it was so terrible that even boxes for fish rattled in air-holes cut in a frozen pond...*

Gorzów Śl. (Landsberg), Poland

1. Schlesische Provinzialblätter, 1786, p. 595.
Gorzów Śl. is only mentioned that there tremors were felt but not as much as in Brzeg, Nysa, Koźle, Racibórz, Prudnik, Krapkowice, Głogówek, Opole, Lewin Brzeski, Stolec and Ziębice.

Gręboszów (Gremboszow), Poland

1. Bójko, 1911, p. 49
...Of interest is a note from the year 1786. Priest Ankiewicz reports about an earthquake at ***Gręboszów*** *which occurred on December 3, the first Sunday of the Advent, at 8 p.m. and was so strong that buildings creaked...*

Iwanowice, Poland

1. Wojciechowski, 1895, p. 323
*…And at **Iwanowice** two miles from Kalisz at the same time, the walls at the old castle were trongly shaken. Of this I am certain, as I was told by a reliable person, who was there…*

Jaksice, Poland

1. Uście Solne - manuscript
*…They said that in some places this earthquake had lasted longer; at **Jaksice** within three quarters of an hour it was repeated three times…*
see: Uście Solne

Janów, Poland

1. Kraków AKMK-11, manuscript]:
*…At **Janów** it happened that doors of rooms opened by themselves, not harmfully, but threateningly…*

Kalisz, Poland

1. Wojciechowski, 1895, p. 323
…On December 3 in the College … at 5 p.m. …in four rooms in the upper and lower floor …the inhabitants felt a great shaking of walls and furniture … the stove and doors creaked much, and the servant leaning against the wall felt the wall sway and bend in both directions. Father Canon Lubieński, on the upper floor, who experienced sickness, felt a chair rocking underneath him.…This rocking was from south to north, lasting 10 seconds. A similar shaking, the creaking of closed doors and the shaking of open ones were felt in another two rooms on the upper and lower floors … In the town it could be hardly felt, so that nobody noticed it in the whole of the town, apart from the people at the College and the Franciscan Monastery, who at that time were in the gallery, …, and as a response to the sudden tremors, ran away… I am certain of the case of the Franciscan nuns since I heard it from them…

Kalwaria, Poland

1. Gazeta Warszawska, 1786, Dec. 16, No. 100
…at Kalwaria chimneys fell down…

2. Gazeta Wileńska, 1786, Dec. 23, No. 51
The same repeated after Gazeta Warszawska, 1786, Dec. 16, No. 100

Karwina (Karviná), Czech Republik

1. Kárník et al., 1958
Kárník et al. have estimated intensity 6 at Karwina without giving any details.

Kęty, Poland

1. Kęty MO manuscript
…1786, December 3, at six p.m. the third tremor here was stronger, longer and most frightening … The earth shook in an extraordinary way, walls moved in

houses, ceilings creaked, windows rattled, and panes fell from weaker ones, roofs inclined from one side to the other, planks and gables fell, and hens dropped from their perches. Truly, there came on the Earth the great fright of the final going away. There was also a great crash, as if number of heavily loaded cars had ran amok making a great noise. In churches ceilings fell and walls broke apart. At the Kęty Town Hall the clock went out of tune. In another area, oaks were broken to half their height, and in forests trees were felled - this earthquake took three prayer's time...

The note was probably written in March 1795, eight years after the earthquake, by J. Kotoński, major of Kęty.

2. Grabowski, 1900, p. 188
mention

3. Grabowski, 1909, p. 182–183
... *The earth shook strongly it seemed that plough-hands were entering the farm-yard with a log of wood... Exactly then came an enormous noise as if the wall were collapsing... the sisters knelt and crying began to pray... The kitchen took fire... the brick fireplace collapsed as it was not supported by a wall or a ceiling...*

4. Droździk, 1979, p. 84.

the same after Kęty MO manuscript

Kielce, Poland

1. Gazeta Warszawska, 1786, Dec. 16, No. 100
Kielce is only mentioned that there the earthquake was felt.

2. Gazeta Wileńska, 1786, Dec. 23, No. 51
The same repeated after Gazeta Warszawska, 1786, Dec. 16, No. 100

3. Lwowskie Tygodniowe Wiadomości, 1786, Dec. 27, No. 96
mention probably after Gazeta Warszawska, 1786, Dec. 16, No. 100

4. Carosi, 1789, p. 37
It is only mentioned that there the earthquake was felt.

Kije, Poland

1. Kraków AKMK-6, manuscript
... *In the whole Decanate the tremor was described differently in various parishes...*

The Kije Decanate comprised parishes of Gnojno, Lisów, Kije, Korytnica and Pińczów.

Kluczbork (Kreuzburg), Poland

1. Schlesische Provinzialblätter, 1786, p. 595
Kluczbork is only mentioned; tremors were felt there, but to a lesser extent than in Brzeg, Nysa, Koźle, Racibórz, Prudnik, Krapkowice, Głogówek, Opole, Lewin Brzeski, Stolec and Ziębice.

Końskie, Poland
1. Carosi, 1789, p. 37
...Farther away in the mountains and iron mines, like Radoszyce, Końskie, Drzewica etc. *nothing was felt...*

Koprzywnica, Poland
1.Kraków AKMK-7, manuscript
....The earthquake greatly frightened the inhabitants. It lasted here no longer than one minute. This tremor surprised all those unemployed and not having fun; in all houses and at all gatherings most of talking is about this tremor...

Korytnica, Poland
1. Kraków AKMK-6, manuscript
...The Korytnica parish reported that the earthquake at the manor of His Excellency Krajczy was so strong that attended guests out of their surprise were unable to talk to one another, except one who called for rescue, as tremor was heavily felt, doors and windows rattled...

Košice (Koszyce, Kaschau, Kassan), Slovakia

1. Brünner Zeitung, 1786, No. 101
...The earthquake was felt at Košice, Iglo and Kirchdorf at about the same time as at Pszczyna...

2. Jeitteles, 1860, p. 329
mention after Brünner Zeitung

3. Kárník et al., 1958, p. 444
Košice is only mentioned

4. Martinka, 1927, p. 152
December 3, 1786 – there was an earthquake at Žilina, ***Košice****, Spišská Nová Ves, Spišske Podhradie with 3 tremors that caused chimneys fall down*

Kowala, Poland

1. Gazeta Warszawska, 1786, Dec. 23, No. 102
...At Kowala... with noise like a carriage running on a cobbled street, the local palace and other masonry and wooden houses shook greatly; the people standing close to one another were jostled because of the great earthquake; pots on stoves in some houses fell to the ground...

2. Lwowskie Tygodniowe Wiadomości, 1787, Jan. 3, No. 98
the same after Gazeta Warszawska

Koźle (Cosel), Poland

1. Schlesische Provinzialblätter, 1786, p.595
Koźle is only mentioned that there the earthquake was strong but not as much as at Bytom, Tarnowskie Góry, Świerklaniec and Pszczyna.

Kraków (Cracow), Poland

1. Trzciński, 1787, p. 26–27
...On December 3, after 05 p.m., again the second time this year the earthquake was felt in ***Kraków****, the direction - ran along the line mean between the summer sunrise and midnight. This tremor was far more violent and longer than the first one* (i.e., on February 27, 1787)..... *at my accommodation... suddenly I heard this sepulchral noise and... immediately I felt the movement of the ground in keeping with the aforesaid direction... . With great fear I noticed that at that moment the top of the wall inclined towards the direction of the movement. This inclination of the wall... showed me the line of the direction. This shaking of the wall in both directions took longer than five heartbeats...*

2. Kraków AP-1, manuscript
...In ***Kraków*** *and the surrounded areas a great earthquake came for the third time this year, but the earlier two were not so felt... whereas this one which happened at quarter past 5, most greatly frightening and saddening everyone. Because of the earthquake, the vaulting of our church cracked in six spots over the chancel, disintegrating with great danger, which was quickly repaired...*

3. Kraków AKK, manuscript
...The second much more violent ... so that statues in different buildings came down as, e.g., at Kazimierz, ***Kraków*** *in the Lateran Canons Corpus Christi Church. A greater shake was felt in St Catherine's Church of the Augustinian Friars, where the vaulting broke, opening wide and threatening to collapse...*

4. Kraków AKAP-2, manuscript
... The earthquake was so much notable and dangerous because in some spots falling stones destroyed the walls of the Augustinian church, and the waters of the Vistula River rose and flooded; different signs of destruction remained in many houses... It lasted no longer than 5 s...

5. Kraków AREF, manuscript
... A noteworthy earthquake, which lasted very short, as if half of an Ave Maria, was observed by the whole congregation, meditating in the refectory and hearing the table move - the same was noticed by other friars having to stay in their cells...

6. Wrocław BOS, manuscript
... A very strong earthquake was felt throughout ***Kraków****, with noise that I experienced myself. Sitting near the wall, I was suddenly pushed away on my stool, and it seemed to me that all the walls would collapse, but I never heard of any harm to anyone...*

7. Gazeta Warszawska, 1786, Dec. 09, No. 98, supplement
...Last Sunday at five past five in the afternoon, throughout the city and its environs an earthquake was felt, lasting several seconds, longer and stronger than the one in February however, without any greater harm, apart from some small cracks. The shaking of walls, doors, windows and ceilings in all houses and

edifices increased more and more, and the final shaking was quite strong and violent… In the course of the tremor itself the noise was heard in many houses…

8. Gazeta Warszawska, 1786, Dec. 13, No. 99
…On December 3 at five past five in the evening a great noise, as if there were carriages speeding on cobbles, was heard in all the streets of ***Kraków****. At once an earthquake was felt, beginning in Kazimierz to the north, namely by swinging… The swinging to the north was greater than the returning to the south. So all the houses began to sway, like a boat on water. Roofs, beams and joints began to creak. All doors and windows rattled. Cupboards and walls themselves inclined distinctly. Stools and chairs under those seated gave way from the south to the north. The floors and ceilings seemed to bend. At the Clock Tower the hammers of the clock made so much noise that not only the officer but other frightened soldiers ran away from the Guardhouse. Also at the tower of the Church of the Most Blessed Saint Mary, the trumpeter heard a strong noise; he assumed it was made by an abrupt entrance of a thief and run to meet him with an axe. Then he noticed that the bell together with its support was moving by itself and he got so scared that half-dead escaped. Some monks present at that time in the refectory felt such a fear that they run away from this place…At the same time the Vistula seethed enormously and rose an ell, so that the people crossing the bridge floating on the river thought that it was sinking underneath them…*

9. Gazeta Warszawska, 1786, Dec. 16, No. 100
… In ***Kraków*** *at many places walls cracked, in particular at the Monastery of the Carmelites at Piasek. In this monastery one of the friars felt very weak in the legs; but when the earth shook and the wall in his cell cracked over the door… he escaped from the cell by himself…The huge and high church of Augustian Friars got so cracked in the roof of its presbytery that it had to be closed for safety reasons…*

10. Gazeta Wileńska, 1786, Dec. 16, No. 50, supplement
The same repeated after Gazeta Warszawska, 1786, Dec. 09, No. 98, supplement

11. Gazeta Wileńska, 1786, Dec. 23, No 51
The same repeated after Gazeta Warszawska, 1786, Dec. 13, No. 99 and Dec. 16, No. 100

12. Lwowskie Tygodniowe Wiadomości, 1786; Dec. 20, No. 94; Dec. 23, No. 95; Dec. 27, No. 96
referred after Gazeta Warszawska

13. Schlesische Privilegirte Zeitung, 1786, Dec. 11, No. 146
…The news came from Cracow that the earthquake lasted two seconds, but … caused no damage. According to the tower guard, it was stronger than the previous one, for the tower moved so strongly that the clock struck twice…

14. Schlesische Privilegirte Zeitung, 1786, Dec. 20, No. 150 and Dec. 23, No. 151
referred after Gazeta Warszawska, 1786, No 99, with slight exaggeration of the events and erroneous interpretation of 'Kazimierz' - a district of Cracow as another old Polish town Kazimierz by the Vistula river.

15. Brünner Zeitung, 1786, No. 101
referred after Schlesische Privilegirte Zeitung, No, 146 and No. 150

16. Pamiętnik Historyczno Polityczny 1787, vol. IV, 237–238.
*...Even in **Kraków** and Pińczów smalller bells swayed so heavily that they rang on their own...*

17. Krzyżanowski, 1900, p. 29–30
*... terrifying roar and underground rumble was heard and the whole of **Kraków** was shaken several times; the earthquake was shorter than that on 27 February, but far more violent and appaling ... nothing collapsed anywhere, but in many places walls and ceilings cracked ...*

18. Carosi, 1789, p. 35–36
...On December 3 at 5 08 p.m. we heard strong underground noise, it lasted 5 seconds and immediately afterwards it came the earthquake from the south towards north. They were not distinctly separated tremors, but rather quite strong swaying in two determinate directions, lasting about 10 s. This was the strongest earthquake, which occurred here from immemorial time but it caused no noticeable damage. Cups, glassware etc. moved on tables, chests and wardrobes; paintings, birdcages,that were fastened to walls swayed and some fell to the ground; open doors closed, the ajared ones opened and these well closed, creaked and trembled. Also, a few pieces of a cracked cornice fell to the ground. Many people observed that walls in rooms inclined and straightened, others noticed this phenomenon for chimneys and roofs. There were places, where the earthquake was felt stronger than anywhere else (where layers of rocks are near the surface), standing persons found it difficult to stay on their feet and there were even those who fell over. ... Raftsmen saw that the Vistula swelled and felt very distinctly their boats sway ... The adjacent water-meadows covered by ice, cracked much with terrifying noise... It is the third earthquake here during 14 months. Phenomenon was not known to ancestors, as it occurred at very long time intervals...

19. Hoff, 1841, p. 78
mentioned

20. Grabowski, 1900, p. 188
...the vaulting of this church (St Catherine's of the Augustinians) in the prima navi was so cracked that because of the danger it had to be closed ...

21. Grabowski, 1909, p. 182
mentioned

22. Kraków AKMK-1, manuscript
The text reflects the mood of awe after the earthquake. It does not give any macroseismic data.

23. Kraków AKMK-2, manuscript
The text reflects the mood of awe after the earthquake. It does not give any macroseismic data.

Jeitteles, 1860; Laska, 1902; Sieberg, 1940; Réthly, 1952; Kárník et al., 1958; Pagaczewski, 1964; Pagaczewski et al., 1972; Kaleta, 1964; and Olczak, 1966a, b, gave also more or less ample information, taken from the sources cited above.

Krapkowice (Krappitz), Poland

1. Schlesische Provinzialblätter, 1786, p. 595
Krapkowice is only mentioned that there the earthquake was strong but not as much as at Bytom, Tarnowskie Góry, Świerklaniec and Pszczyna.

Kromołów, Poland

1. Kraków AKMK-9, manuscript
…On December 3 after 5 p.m. a considerable earthquake was felt here, to which I was quite sensitive myself, when the walls of the whole wooden building shook. Glasses in the cupboard rattled, doors opened by their own, the legs of the table I was sitting by shook like on a swamp. In the town and the villages belonging to the parish similar effects could be felt. The water in the pipes in the distillery splashed to the ground, the pots in cupboards and hens on their roosts fell to the ground, and a great underground noise could be heard… The tremor caused no … damage…

Krzeszowice, Poland

1. Gazeta Warszawska, 1786, Dec. 13, No. 99
*… The church at **Krzeszowice** shook so heavily that the bell hung in the tower was heard ringing…*

2. Gazeta Wileńska, 1786, Dec. 23, No. 51
The same repeated after Gazeta Warszawska, 1786, Dec. 13, No. 99

3. Lwowskie Tygodniowe Wiadomości, 1786, Dec. 23, No. 95
Krzeszowice is only mentioned that there the earthquake was felt and heard very much.

4. Schlesische Privilegirte Zeitung, 1786, Dec. 20, No. 150
information after Gazeta Warszawska but much exaggerated:
*… in **Krzeszowice** the church was damaged and the bells tolled very heavily…*

5. Brünner Zeitung, 1786, No. 104
after Schlesische Privilegirte Zeitung, 1786, Dec. 20, No. 150

6. Jeitteles, 1860, p. 329
after Brünner Zeitung, 1786, No. 104

Kurozwęki, Poland

1. Gazeta Warszawska, 1786, Dec. 23, No. 102
*…At **Kurozwęki**…the closed doors in the palace rooms suddenly opened. Stools and tables moved visibly, papers on tables rustled, keys hanging at clocks rang. The wall wainscoting clattered, mirrors moved* etc., *greatly terrifying those present …*

2. Lwowskie Tygodniowe Wiadomości, 1787, Jan. 3, No. 98
mentioned after Gazeta Warszawska

Legnica (Liegnitz), Poland

1. Schlesische Provinzialblätter, 1786, p. 595
... In Wrocław ...most people only noticed very slight shaking of doors and that cups and glasses rattled. The same was noticed in ***Legnica*** *and Głogów...*

Lewin Brzeski (Lowen), Poland

1. Schlesische Provinzialblätter, 1786, p. 595
Lewin Brzeski is only mentioned that there the earthquake was strong but not as much as at Bytom, Tarnowskie Góry, Świerklaniec, and Pszczyna.

Lipowiec, Poland

1. Gazeta Warszawska, 1786, Dec. 13, No. 99
... It is 15 miles from ***Lipowiec*** *to Opatów and in both places it was strongly felt..*

(1 Polish mile = 7146 m)
2. Gazeta Wileńska, 1786, Dec. 23, No. 51
The same repeated after Gazeta Warszawska, 1786, Dec. 13, No. 99

Lisów, Poland

1. Kraków AKMK-6, manuscript
...The ***Lisów*** *parish reported that in some villages it was stronger, but in others weaker, and that was the reason why at the presbytery, only those paying attention, noticed that it trembled. And in addition to that report, in the manor house of village Brudzów, belonging to the same parish, an earthquake was felt again at 8 o'clock at sunrise on December 9. Hooks on the doors were seen to move, locks were detached, and bars hit against window panes. Every movement of the earth lasted long enough to say 5 or 6 words. Also, from other parishes came reports that stronger or weaker earthquakes were felt there, but nothing important is to be added to the above description....*

Litovel (Littau), Czech Republic

1. Brünner Zeitung, 1786, No. 102
... According to a report from ***Litovel,*** *it was so strong that, among other things, several vaultings cracked considerably...*

2. Jeitteles, 1860, p. 329–330
after Brünner Zeitung, 1786, No. 102

3. Remes, 1902, p. 8
after Brünner Zeitung, 1786, No. 102

4. Kárník et al., 1958, p. 444
Kárník et al. have estimated intensity 5-7 without giving any details.

Lubsza (Lubschau), Poland

1. Schlesische Privilegirte Zeitung, 1786, Dec. 11, No. 146
... We still have little information from outside [Tarnowskie Góry], *except that it was felt in **Lubsza** and Świerklaniec.*

Lwów (Lemberg, Lvov), Ukraine

1. Brünner Zeitung, 1786, No. 103, p. 817
... In our capital, though hardly noticeable, it was nevertheless felt by different people in rooms where it was very quiet...

2. Jeitteles, 1860, p. 329
...weakly felt also in Lwów

3. Laska, 1902, p. 30
mention probably after Brünner Zeitung, 1786, No. 103, p. 817

Małogoszcz, Poland

1. Gazeta Warszawska, 1786, Dec. 16, No. 100
*...The earthquake reached only **Małogoszcz**, for beyond this place, towards Warszawa, it was not felt...*

2. Gazeta Wileńska, 1786, Dec. 23, No. 51
The same repeated after Gazeta Warszawska, 1786, Dec. 16, No. 100

3. Lwowskie Tygodniowe Wiadomości, 1786, Dec. 27, No. 96
after Gazeta Warszawska, Dec. 16, No. 100

4. Carosi, 1789, p. 37
It is only mentioned that the earthquake was felt "beyond Małogoszcz".

Milicz (Militsch), Poland

1. Schlesische Provinzialblätter, 1786, p. 595
*...It was stronger in Bierutów (than in Oleśnica), however it was far weaker in area of Dobroszyce, Trzebnica, Twardogóra, **Milicz**, Sułów, Zduny, and Rawicz. In most of these localities - apart from the fear of those who experienced it - it was manifested by a slight shaking of chairs, tables, windows in rooms, and all standing objects. Also at this occasion, it was noticed that birds and other animals showed the same signs of fear as those in February...*

Mogiła, Poland

1. Kraków-Mogiła AOC, manuscript
... during a dinner there was a considerable tremor . Fear seized everyone, who in awe, left their place, ... especially the teacher, who ran away. So great was the force, so great the movement, that even the bell hanging in the gateway rang and the paintings hung on the walls suddenly fell off...

Mrzygłód, Poland

1. Kraków AKMK-10, manuscript
...On December 3 at 05:15 p.m. an earthquake was felt; it took no longer than 1-2 minutes. There came tremendous noise as if there was a most violent wind, and then we could hear the greatest roars coming, moving tables, chairs and glasses, and cups in the sideboard. Moreover, walls and roofs began to creak, and in some town houses pots and household implements began to move. The doors in rooms and sties opened so that some people ran away outdoors and others did not know what to do. In other villages belonging to this parish, towards Siewierz, the people heard underground rumbles, as if guns had been fired, and all the houses and cottages were shaken up. The doors opened, window panes broke, pots and other vessels fell from the shelves, and the people thought that the earth was falling down with them...

Myślenice, Poland

1. Kraków AP-3, manuscript
... The earthquake lasted 2 minutes and was quite strong, since in some houses chairs, benches and other furniture overturned...

2. Carosi, 1789, p. 37
...The same underground noise - which preceeded the earthquake - and the next shakes were felt in Tarnów, ***Myślenice****, Zator, Oświęcim, Bielsko-Biała, and many other places...*

3. Kreutz, 1876, p. 170
mention

Nadzów, Poland

1. Gazeta Warszawska, 1786, Dec. 9, No. 98, supplement
... The news received from ***Nadzów****, Pińczów, Wawrzenczyce and many other localities in the provinces of Kraków and Sandomierz ensure that both provinces felt the effects of this earthquake ...*

2. Gazeta Wileńska, 1786, Dec. 16, No. 50, supplement
The same repeated after Gazeta Warszawska, 1786, Dec. 9, No. 98, supplement

3. Lwowskie Tygodniowe Wiadomości, 1786, Dec. 20, No. 94
after Gazeta Warszawska, Dec. 9, No. 98, supplement

Namysłów (Namslau), Poland

1. Schlesische Privilegirte Zeitung, 1786, Dec. 9, No. 145
*...**Namysłów**, December 4... our city experienced a natural event yesterday evening, which terrified local residents. At 4.56 p.m. a strong tremor shook the local buildings, in particular those situated in the eastern part of the town, so hard that the bell of the clock on the town hall tower struck 6-7 times, and tin vessels moved on shelves in roos. In many houses open doors in cupboards swayed, and chairs moved from their places ...A cage moved so much that the bird fell from its roost ...It was noticed that the bridge swayed strongly, and many inhabitants heard the strong underground rumble...*

2. Schlesische Provinzialblätter, 1786, p. 595
...The fact that the hammer struck the bell was noticed also in Namysłów and several other localities...

3. Gazeta Warszawska, 1786, Dec. 30, No. 104
after Schlesische Privilegirte Zeitung

4. Hoff, 1841, p. 78
mentioned

5. Jeitteles, 1860, p. 330
mention after Schlesische Privilegirte Zeitung

Nowa Góra, Poland

1. Kraków AKMK-12, manuscript
... The previous year in the parish, at the time after the earthquake in the month of December, there were many diseases, none of which however was fatal...

Nowy Sącz, Poland

1. Kraków AP-3, manuscript
...We should inform about extraordinary events happened in the district, i.e. about the earthquake felt throughout the whole country on 3 of the current month between 5 and 6 p.m., probably from the west... It caused no visible damage...

2. Kreutz, 1876
mention after Kraków AP-3

Nowy Targ (Neumark), Poland

1. Brünner Zeitung, 1786, No. 103, 817–818
... the earthquake affected our ***Nowy Targ*** *to a greater extent, higher than elsewhere . This day on afternoon between 5 and 6 for a few minutes, a terrifying underground rumble was heard, then one after the other two tremors followed; and about 5 minutes later came the third one, most violent, causing deadly fear among the inhabitants. It was so strong that - as was noticed later - rocks in the mountains cracked at a boot's width apart ... almost in all houses, stoves collapsed. The only two stone houses cracked from the foundation to the roof so much that a man's thumb could be stuck in these fissures. The wooden houses which were poorly set in foundations were moved and twisted so that the people will have to work in order to put their dwellings back in order...*

2. Jeitteles, 1860, 329–330
the same after Brünner Zeitung, No. 103, 817–818

Nysa (Neise), Poland

1. Schlesische Privilegirte Zeitung, 1786, Dec. 9, No. 145
...In Brzeg, ***Nysa*** *and Głubczyce the earthquake was stronger...(than in Wrocław).*

2. Gazeta Warszawska, 1786, Dec. 30, No. 104
the same after Schlesische Privilegirte Zeitung

3. Schlesische Provinzialblätter, 1786, p. 595
Nysa is only mentioned that there the earthquake was strong but not as strong as at Bytom, Tarnowskie Góry, Świerklaniec and Pszczyna.

4. Hoff, 1841, p. 78
mentioned

5. Jeitteles, 1860, p. 330
mention after Schlesische Privilegirte Zeitung

Oleśno (Rosenberg), Poland

1. Schlesische Provinzialblätter, 1786, p. 595
Oleśno is only mentioned; tremors were felt but to a lesser extent than in Brzeg, Nysa, Koźle, Racibórz, Prudnik, Krapkowice, Głogówek, Opole, Lewin Brzeski, Stolec and Ziębice.

Oleśnica (Oels), Poland

1. Schlesische Provinzialblätter, 1786, p. 595
... This earthquake was also noticed in ***Oleśnica****, at the ducal castle and several places in the town...*

Olkusz, Poland

1. Gazeta Warszawska, 1786, Dec. 13, No. 99
...At Olkusz over several seconds, the earth moved so strongly thrice that chairs and stools swayed in rooms and frightened people dropped shouting to the ground...

2. Gazeta Wileńska, 1786, Dec. 23, No. 51
The same repeated after Gazeta Warszawska, 1786, Dec. 13, No. 99

3. Schlesische Privilegirte Zeitung, 1786, Dec. 20, No. 150
after Gazeta Warszawska, slightly exaggerated

4. Lwowskie Tygodniowe Wiadomości, 1786, Dec. 23, No. 95
Olkusz is only mentioned that there the earthquake was heard and felt strongly.

5. Brünner Zeitung, 1786, No. 104, p. 826
mention after Schlesische Privilegirte Zeitung

6. Jeittles 1860, p. 329
mention after Brünner Zeitung

Olomouc (Ollmütz, Ołomuniec), Czech Republic

1. Brünner Zeitung, 1786, No. 104
... It could be felt, in particular in the western part of the town, where some inhabitants left their homes and run out into the street...

2. Jeitteles, 1860, p. 329–330
after Brünner Zeitung

3. Remes, 1902, p. 8
after Brünner Zeitung

4. Kárník et al., 1958, p. 444
Kárník et al. have estimated intensity 5 at Olomouc without giving any details.

Opava (Opawa, Troppau), Czech Republic

1. Gazeta Warszawska, 1786, Dec. 16, No. 100

2. Gazeta Wileńska, 1786, Dec. 23, No. 51
The same repeated after Gazeta Warszawska, 1786, Dec. 16, No. 100
*...Came somebody from abroad to Kraków and reported about a felt earthquake at **Opawa**...*

3. Jeitteles, 1860, p. 330
*... In **Opava** too the earthquake turned out to be really great and still lingers in reminiscences of older people...*

Opatów, Poland

1. Gazeta Warszawska, 1786, Dec. 13, No. 99
*... It is 15 miles from Lipowiec to **Opatów**, and in both places it was strongly felt ...* (1 Polish mile = 7146 m)

2. Gazeta Wileńska, 1786, Dec. 23, No. 51
The same repeated after Gazeta Warszawska, 1786, Dec. 13, No. 99

3. Schlesische Privilegirte Zeitung, 1786, Dec. 20, No. 150
mention after Gazeta Warszawska

Opole (Oppeln), Poland

1. Schlesische Provinzialblätter, 1786, p. 595
Opole is only mentioned that there the earthquake was strong but not as much as at Bytom, Tarnowskie Góry, Świerklaniec and Pszczyna.

Oświęcim, Poland

1. Gazeta Warszawska, 1786, Dec. 16, No. 100
*...There was also the earthquake in Busko, Kielce, Chęciny, Głębowice and the whole of **Oświęcim** and Zator Duchies and closer to Kraków than stronger...*

2. Gazeta Wileńska, 1786, Dec. 23, No. 51
The same repeated after Gazeta Warszawska, 1786, Dec. 16, No. 100

3. Carosi, 1789, p. 37
*...The same underground noise - which preceded the earthquake - and the next shakes were felt in Tarnów, Myślenice, Zator, **Oświęcim**, Bielsko-Biała, and many other places...*

Pińczów, Poland

1. Kraków AKMK-6, manuscript
*… The **Pińczów** parish also reports on a considerable earthquake, prior to which a rumble was heard, like men on horseback, causing the ave-bell to be heard at the church…*

2. Gazeta Warszawska, 1786, Dec. 9, No. 98, supplement
*…The news received from **Pińczów**, Wawrzeńczyce, Nadzów and many other localities in the provinces of Kraków and Sandomierz ensure that both provinces felt the effects of this earthquake…*

3. Gazeta Wileńska, 1786, Dec. 16, No. 50, supplement
The same repeated after Gazeta Warszawska, 1786, Dec. 9, No. 98, supplement

4. Lwowskie Tygodniowe Wiadomości, 1786, Dec. 20, No. 94
after Gazeta Warszawska,1786, Dec. 9, No. 98, supplement

5. Carosi, 1789, p. 37
*… The extension … from the south to the north, where this event was felt… beginning from Kraków … beyond **Pińczów** and the surrounding areas…*

6. Pamiętnik Historyczno Polityczny, 1787, vol. IV, p. 237–238
*… Even in Kraków and **Pińczów** … smaller bells swayed so much that they rang on their own…*

7. Wojciechowski, 1895, p. 323
…It was really an earthquake felt this day at the same time in Kraków, Pińczów, Tarnów, Sancygniów, Piotrków and other places quite distant from one another…

Piotrków, Poland

1. Gazeta Warszawska, 1786, Dec. 09, No. 98, supplement
*…It is reported from **Piotrków**, that in the town, this evening the earthquake was felt, causing no visible damage…*

2. Gazeta Wileńska, 1786, Dec. 16, 1786, No, 50, supplement
The same after Gazeta Warszawska, 1786, Dec. 09, No. 98, supplement

3. Gazeta Warszawska, 1786, Dec. 16, No. 100
…At the Piarist College, stoves, doors, windows, and even walls were seen moving. One monk who was sitting on a chair was abruptly displaced from south to north… In another cell, the chair with a monk sitting on it and the table kept moving to and fro in various directions. Third monk, who happened to be standing, begun to tremble and had to lean against the table in order not fall down, in doing so he noticed that a ruler hanging on a wall was in a swing - like motion. Forth monk, who was also standing, was shaken and wanted to get support from the wall, but the wall kept thrusting him away. Fifth monk, who was sitting by the stove, was strongly hit by it. In the college some students and the director wanted to adhere to the wall, but the wall was rejecting them and they kept jumping. The

same happened at the Dominican monastery and in other places; but everywhere, thank God, there was no damage…

4. Gazeta Wileńska, 1786, Dec. 23, 1786, No, 51
The same after Gazeta Warszawska, 1786, Dec. 16, No. 100.

5. Kraków AKAP, manuscript
… *This earthquake spread from the north to the south, extending from* ***Piotrków*** *to Puławy and lasting no more than 5 seconds…*

6. Carosi, 1789, p. 37
Piotrków is only mentioned that there the earthquake was felt.

7. Pamiętnik Historyczno Polityczny, 1787, vol. IV, p. 237–238
…The last earthquake …spread farther than the previous one, because it was felt from Tarnów up to ***Piotrków****…*

8. Wojciechowski, 1895, p. 323
…It was really an earthquake felt this day at the same time in Kraków, Pińczów, Tarnów, Sancigniów, ***Piotrków*** *and other places quite distant from one another…*

9. Krzyżanowski, 1900, p. 30
… *at that time this earthquake was felt far from Kraków, even up to Piotrków…*

Pokój (Carlsruhe), Poland

1. Schlesische Provinzialblätter, 1786, p. 595
…In ***Pokój*** *the clock bell hammer struck repeatedly at the castle and at the church, and the eagle on the top of the tower turned continually during this earthquake…*

Polanka (Polonca), Poland

1. Lwowskie Tygodniowe Wiadomości, 1786, Dec. 20, No. 94
…There was a report from ***Polanka*** *in Myślenice area, that there the earthquake was on 3 December at 15 minutes past five in the evening… An underground rumbling, sounding like a heavy carriage running over a rough, lumpy road was heard. At the same moment, the earth began to tremble, at first violently, but only for a short time, and then a slower movement for a good minute. In all the nearby houses, the same was felt, and people saw the doors shake, walls and roofs creaked, but no major damage occurred. It was clearly seen that the quake proceeded from west to east…*

Popędzyna (Popedzyn) Poland

1. Uście Solne - manuscript
The earthquake was felt in Popędzyna as much as in Uście Solne. For detailed description of the event see Uście Solne.

2. Bójko, 1911, p. 49
after Uście Solne - manuscript

Potok Złoty (Potok), Poland

1. Kraków AKMK-11, manuscript
...I send this news that on 3 of this month about 5 at the evening throughout the ***Potok*** *parish the earthquake was felt. True, I did not notice it, because it was not the same in all places ... Whereas in some houses they felt little, and more in others, and in particular in two cottages where hens fell from their roosts ... I have heard that in some fields the earth caved in...*

Prudnik (Neustadt), Poland

1. Schlesische Provinzialblätter, 1786, p. 595
Prudnik is only mentioned that there the earthquake was strong but not as much as at Bytom, Tarnowskie Góry, Świerklaniec, and Pszczyna.

Pszczyna (Pless), Poland

1. Schlesische Privilegirte Zeitung, 1786, Dec. 11, No. 146
...In ***Pszczyna*** *on Sunday, December 3 at 04 45 p.m. here we felt quite a strong earthquake, so that most people ran out in a hurry of their homes: indeed, because some houses cracked and stoves were in ruin so that on the next days rooms could not be heated because of the smoke coming out. Hens and birds fell from roosts, and in some localities glasses fell off. The tremor did not seem to be as strong as that in February, when its effects were greater. At that time the people outdoors concerted in saying that they felt as if had fallen down together with the ground and had then been thrown up...*

2. Schlesische Provinzialblätter, 1786, p. 595
At many houses (the same as in ***Pszczyna*** *and a few places in Poland) walls cracked from top to bottom... In* ***Pszczyna****, too, the effect was almost as violent (as in Tarnowskie Góry), even there many houses cracked, in particular stoves in rooms became so warped that the next day they could not be fired ...*

3. Brünner Zeitung, 1786, No. 101, p. 799
after Schlesische Privilegirte Zeitung, 1786, Dec. 11, No. 146

4. Jeitteles, 1860, p. 329
after Brünner Zeitung, 1786, No. 101, p. 799

Puławy, Poland

1. Kraków AKAP-2, manuscript
... This earthquake spread from the north to the south, extending from Piotrków to ***Puławy*** *and lasting no more than 5 seconds...*

Racibórz (Rattibor), Poland

1. Schlesische Privilegirte Zeitung, 1786, Dec. 09, No. 145
... From a letter from ***Racibórz****: Today at 05:10 p.m. we felt again in* ***Racibórz*** *an earthquake. There was a noise in the air and soon the earthquake lasting*

about 4 seconds from the direction again from the east to the west, however it was not so strong as in February, but it was noticed that an inkpot fell from the table. Also books and newspaper fell off... News from Wrocław: In Racibórz two thick walls broke up and glasses fell off in various houses...

2. Gazeta Warszawska, 1786, Dec. 30, No. 104
after Schles. Priv. Zeit., but exaggerated: ... *some houses cracked...*

3. Schlesische Provinzialblätter, 1786, p. 594
Racibórz is only mentioned that there the earthquake was strong but not as much as at Bytom, Tarnowskie Góry, Świerklaniec and Pszczyna.

4. Hoff, 1841, p. 78
mentioned

5. Jeitteles, 1860, p. 330
mention after Schlesische Privilegirte Zeitung

6. Laska, 1902, p. 29
mention after Gazeta Warszawska, exaggerated - It is a translation from Polish into German after Gazeta Warszawska, failing to draw on the source, i.e., Schlesische Privilegirte Zeitung

Racławice, Poland

1. Gazeta Warszawska, 1786, Dec. 23, No. 102
... *In **Racławice** the sacristy collapsed...*

2. Lwowskie Tygodniowe Wiadomości, 1787, Jan. 03, No. 98
mention after Gazeta Warszawska

Radoszyce, Poland

1. Carosi, 1789, p. 37
...Farther away in the mountains and iron mines, like Radoszyce, Końskie, Drzewica etc. *nothing was felt...*

Rawicz (Rawitsch), Poland

1. Schlesische Provinzialblätter, 1786, p. 595
*...It was stronger in Bierutów (than in Oleśnica), however it was far weaker in area of Dobroszyce, Trzebnica, Twardogóra, Milicz, Sułów, Zduny, and **Rawicz**. In most of these localities - apart from the fear of those who experienced it - it was manifested by the slight shaking of chairs, tables, windows in rooms, and all standing objects. Also at this occasion, it was noticed that birds and other animals showed the same signs of fear as those in February...*

Rybna, Poland

1. Schlesische Privilegirte Zeitung, 1786, Dec. 11, No. 146
*... but in **Rybna** nearby, the tremor was even stronger than in the town ...* i.e. Tarnowskie Góry

Rybnik (Riebnick), Poland

1. Schlesische Privilegirte Zeitung, 1786, Dec. 09, No. 145
... They noticed the same in ***Rybnik*** (see Racibórz). *The guardhouse said that the town hall tower swayed, also different burghers said that the roofs were supposed to creak ... The ice on the water surrounding the Castle creaked in various ways...*

2. Jeitteles 1860, p. 330
mention after Schlesische Privilegirte Zeitung

Sancygniów, Poland

1. Pamiętnik Historyczno Polityczny, 1787, vol. III, p. 237–238
... smaller bells swayed so much that they rang on their own...

2. Wojciechowski, 1895, p. 323
...It was really an earthquake felt this day at the same time in Kraków, Pińczów, Tarnów, ***Sancygniów****, Piotrków and other places quite distant from one another...*

Sandomierz, Poland

1. Kraków AKMK-4, manuscript
... Between 5 and 6 p.m. the whole town was affected by the earthquake, lasting about a minute, however, with no vehemence. ... The sick, lying in bed at the time, felt that their beds went up. I describe it myself ... that, resting on a chair, I could observe: a chair rose from the south and turned to the north, three times, as it turned for the third time I heard the sound of bells or a bell. I noticed that the clock on the tower of the College struck...

2. Kraków AKAP-1, manuscript
... On 3 December in ***Sandomierz*** *and the surrounded areas an earthquake was felt...*

3. Gazeta Warszawska, 1786, Dec. 09, No. 98, supplement
... The news received from Nadzów, Pińczów, Wawrzeńczyce and many other localities of the provinces of Kraków and ***Sandomierz*** *ensure that both provinces felt the effects of this earthquake...*

4. Gazeta Wileńska, 1786, Dec. 16, No. 50, supplement
The same repeated after Gazeta Warszawska, 1786, Dec. 09, No. 98, supplement

5. Gazeta Warszawska, 1786, Dec. 13, No. 99
... In ***Sandomierz****, at the Reformation monastery, cupboards nailed to the walls moved considerably outwards; in the Jurist houses, papers fell from the tables; in many places glasses turned over, and the Vistula also seethed...*

6. Gazeta Wileńska, 1786, Dec. 16, No. 50
The same repeated after Gazeta Warsyawska, 1786, Dec. 13, No. 99

7. Lwowskie Tygodniowe Wiadomości, 1786, Dec. 23, No. 95
Sandomierz is only mentioned; the earthquake could be heard and felt very much.

8. Schlesische Privilegirte Zeitung, 1786, Dec. 20, No. 150
after Gazeta Warszawska, slightly exaggerated

9. Brünner Zeitung, 1786, No. 104
after Schlesische Privilegirte Zeitung and Gazeta Warszawska

10. Jeitteles, 1860, p. 329
mention after Brünner Zeitung

Spišská Nová Ves (Iglo, Neudorf), Slovakia

1. Brünner Zeitung, 1786, No. 101
*...The earthquake was felt at Košice, **Spišská Nová Ves,** and Spišske Podhradie at about the same time as at Pszczyna. In **Spišská Nová Ves,** it is supposed to have been very strong, for the bells in the tower rang several times, so that inhabitants thought that they heard a fire alert...*

2. Jeitteles, 1860, p. 329
after Brünner Zeitung

3. Kárník et al., 1958, p. 444
Kárnik et al. have estimated intensity 5-6 at Spišska Nová Ves without giving any details.

4. Martinka, 1927, p. 152
*December 3, 1786 – there was an earthquake at Žilina, Košice, **Spišská Nová Ves,** Spišske Podhradie with 3 tremors that caused chimneys fell down*

5. Koláček, 1921, p. 8,
mentioned

6. Mallet and Mallet, 1858, p. 15
at ***Spišská Nová Ves*** *bells rang several times*

7. Perrey, 1846, p. 361
mentioned

Spišske Podhradie (Kirchdorf, Kirchdrauf, Szepesvsraljan), Slovakia

1. Brünner Zeitung, 1786, No. 101
*... The earthquake was felt at Košice, Spišska Nová Ves, and **Spišske Podhradie** at about the same time as at Pszczyna.*

2. Jeitteles, 1860, p. 329
mention probably after Brünner Zeitung

3. Kárník et al., 1958, p. 444
Spišske Podhradie is mentioned.

4. Martinka, 1927, p. 152
*December 3, 1786 – there was an earthquake at Žilina, Košice, Spišská Nová Ves, **Spišske Podhradie** with 3 tremors that caused chimneys fell down*

Stolec (Stolz), Poland

1. Schlesische Provinzialblätter, 1786, p. 595
It is only mentioned, that in Stolec the earthquake was strong but not as much as at Bytom, Tarnowskie Góry, Świerklaniec and Pszczyna.

Sułów (Sulau), Poland

1. Schlesische Provinzialblätter, 1786, p. 595
...It was stronger in Bierutów (than in Oleśnica), however it was far weaker in area of Dobroszyce, Trzebnica, Twardogóra, Milicz, ***Sułów****, Zduny, and Rawicz. In most of these localities - apart from the fear of those who experienced it - it was manifested by the slight shaking of chairs, tables, windows in rooms, and all standing objects. Also at this occasion, it was noticed that birds and other animals showed the same signs of fear as those in February...*

Świerklaniec (Neudeck), Poland

1. Schlesische Privilegirte Zeitung, 1786, Dec. 11, No. 146
... We still have little information from outside [Tarnowskie Góry], *except that it was felt in Lubsza, Świerklaniec and all the villages nearby...*

2. Schlesische Provinzialblätter, 1786, p. 594
...In Bytom, ***Świerklaniec*** *and other localities it was experienced stronger than in Tarnowskie Gory itself...*

Tarnowskie Góry (Tarnowitz), Poland

1. Schlesische Privilegirte Zeitung, 1786, Dec. 11, No. 146
... It was 5:15 p.m., the chair on which I was sitting began to move so much under me that it was something which could not only be felt, but also seen.... The walls began to sway, the windows rattled, the ceiling and the floor visibly moved, and a good deal of dust filled the room; we ran to the door, which we could not open at first, because it was stuck by the constant ongoing movement... A brick and more rubble fell on to the stairs. We made haste to get outdoors and in the market we found those who had left their homes ... The tremors came from the southwest ... incomparably stronger than those on 27 February. No house in the market was safe, in one a chimney cornice fell off breaking many kitchen implements. In most houses, chimney cornices were damaged. The workers at smelter nothing noticed, but in Rybna nearby, the tremor was even stronger than in the town. The earthquake lasted 8-10 seconds, and during that time 3 tremors came...

2. Schlesische Provinzialblätter, 1786, p. 594–595
... The direction of the tremors was the same as that on February 27, and the greatest force also came near ***Tarnowskie Góry****, Pszczyna, Racibórz, Kraków... In Tarnowskie Góry, many inhabitants left their homes and run into the market. In many houses (likewise in Pszczyna and other places of Poland), walls got strong cracks from top to bottom. In one house a cup from the chimney fell off, breaking many kitchen implements. In many other houses cornices partly fell from chimneys*

or strongly cracked. The creaking of beams, the rattling of window panes, the shaking of the doors and the falling debris and brick caused even greater fear among the inhabitants here and the localities nearby as in Bytom and Świerklaniec where the earthquake was stronger than in ***Tarnowskie Góry*** *itself...*

3. Brünner Zeitung, 1786, No. 101
after Schlesische Privilegirte Zeitung

4. Jeitteles, 1860, p. 329
mention after Brünner Zeitung

5. Mallet and Mallet 1858, p. 15
It is mentioned that in Tarnowskie Góry and some other places there were three shocks from the southwest.

Tarnów, Poland

1. Kraków AP-3, manuscript
... In the whole area around here quite a strong earthquake was felt ... In the town of ***Tarnów*** *it caused cracks of walls in 8 houses; ... in 5 houses these cracks were insignificant, but in the remaining three they could become dangerous. Depending on the state, necessary actions were undertaken to prevent any danger...*

2. Tarnów MO, manuscript
...On December 3 at 05:11 p.m. a terrible earthquake occurred in Tarnów that lasted 1 minute. Some houses got cracks and the bells rang twice at St. Catherine's and the church of Bernardine Friars...

3. Tarnów AD, manuscript
The Bishop of Tarnów commanded all lay and order priests to hold public prayers because of the earthquake.

4. Brünner Zeitung, 1786, No. 99
...commonly we felt here an earthquake, fortunately, it caused no damage...

5. Brünner Zeitung, 1786, No. 103
... In letters from ***Tarnów*** *the earthquake is described as very strong. Different things were scattered and the inhabitants ran out of their homes lamenting and fearing that everything may collapse. The town hall bell rang several times...The earthquake, lasted 2 seconds, terrified the inhabitants... They saw that the direction was from north to south ... In some places there was damage although rather slight, but some cracks on the walls are noticeable...*

6. Carosi, 1789, p. 37
...The same underground noise - which preceded the earthquake - and the next shakes were felt in ***Tarnów****, Myślenice, Zator, Oświęcim, Bielsko-Biała, and many other places...*

7. Pamiętnik Historyczno Polityczny, 1787, vol. IV, p. 237–238
... The last earthquake ... spread farther than the previous one, because it was felt from ***Tarnów*** *up to Piotrków...*

8. Kreutz, 1876
*...The earthquake was so strong in **Tarnów**, several houses threatened to collapse because their walls cracked...*
Information from a report to the administration of the Austro-Hungarian province - but slightly exaggerated.

9. Wojciechowski, 1895, p. 323
mention

10. Jeitteles, 1860, p. 329
mention after Brünner Zeitung

11. Chrościński, 1896, p. 79
The earthquake lasted 2 seconds, terrified inhabitants....its direction was from the north to the south. ...it caused unsignificant damage to walls, but in some places, traces of damage are visible

Trzebinia, Poland

1. Kraków AKMK-13, manuscript
They were hold services and prayers in the Trzebinia parish on the occasion of the earthquake.

Trzebnica (Trebnitz), Poland

1. Schlesische Provinzialblätter, 1786, p. 595
*...It was stronger in Bierutów (than in Oleśnica), however it was far weaker in area of Dobroszyce, **Trzebnica,** Twardogóra, Milicz, Sułów, Zduny, and Rawicz. In most of these localities - apart from the fear of those who experienced it - it was manifested by the slight shaking of chairs, tables, windows in rooms, and all standing objects. Also at this occasion, it was noticed that birds and other animals showed the same signs of fear as those in February...*

Twardogóra (Festenberg), Poland

1. Schlesische Provinzialblätter, 1786, p. 595
*...It was stronger in Bierutów (than in Oleśnica), however it was far weaker in area of Dobroszyce, Trzebnica, **Twardogóra,** Milicz, Sułów, Zduny, and Rawicz. In most of these localities - apart from the fear of those who experienced it - it was manifested by the slight shaking of chairs, tables, windows in rooms, and all standing objects. Also at this occasion, it was noticed that birds and other animals showed the same signs of fear as those in February...*

Tyniec, Poland

1. Gazeta Warszawska, 1786, Dec. 13, No. 99
*... At **Tyniec** the church is said to have been cracked...*

2. Gazeta Wileńska, 1786, Dec. 23, No. 51
The same repeated after Gazeta Warszawska, 1786, Dec. 13, No. 99

3. Lwowskie Tygodniowe Wiadomiości, 1786, Dec. 23, No. 95
Tyniec is only mentioned that there the earthquake could be heard and felt very much ...

4. Schlesische Privilegirte Zeitung, 1786, Dec. 20, No. 150
*... in **Tyniec**, too, it was very strong...*
information from Warsaw of 13 December, probably after GazetaWarszawska

5. Brünner Zeitung, 1786, No. 104
after Schlesische Privilegirte Zeitung

6. Jeitteles, 1860, p. 329
mention after Brünner Zeitung

Uście Solne, Poland

1. Uście Solne - manuscript
...On 3 December 1786 at five in the evening after sunset, the earthquake was felt in the town so strongly, that in some houses door locks failed, doors opened, roofs creaked, walls seemed to collapse, and windows and stoves were seen moving so that this event extraordinary here though very short, lasted only one minute, and unnoticed by others, caused much fear and crying. It happened all over the area, at Barczków, Popędzyna, Wrzępia, Bratucice, Bogucice, Cerekiew, Bieńkowice etc., *and was distincly felt by all people, old and young. They said that in some places this earthquake had taken longer, at Jaksice within three quarters of an hour it was repeated thrice. It was started in Bochnia a half an hour before sunset...*

2. Bójko, 1911, p. 49
referred after Uście Solne - manuscript

Wawrzeńczyce, Poland

1. Gazeta Warszawska, 1786, Dec. 09, No. 98, supplement
*...The news received from Pińczów, Nadzów, **Wawrzeńczyce** and many other localities in the provinces of Kraków and Sandomierz ensure that both provinces felt the effects of this earthquake...*

2. Gazeta Wileńska, 1786, Dec. 16, No. 50, supplement
The same repeated after Gazeta Warszawska, 1786, Dec. 9, No. 98, supplement

3. Lwowskie Tygodniowe Wiadomosci, 1786, Dec. 20, No. 94
The same after Gazeta Warszawska, 1786, Dec. 09, No. 98, supplement

Widuchowa, Poland

1. Kraków AKMK-6, manuscript
*... They say that in the whole Gnojno parish there was a great earthquake, that children fell from stove-corners, and that the stoves cracked and the tops of buildings creaked. And even poultry fell from their perches. At **Widuchowa** of the Gnojno parish it was so terrible that even boxes for fish rattled in air-holes cut in a frozen pond...*

Wieliczka, Poland

1. Carosi, 1789, p. 37
... In ***Wieliczka*** *the whole town was afraid and in the salt mine a large block broke away and felt, but houses were not damaged ...*
In the archival materials of the Salt Mine no data on this earthquake have been found; so it may be assumed that the tremor had a slight effect, not of destroying nature.

Wielkanoc, Poland

1. Gazeta Warszawska, 1786, Dec. 13, No. 99
... In the village of ***Wielkanoc*** *when somebody was counting money on the table during this tremor, the money started to split out and there was no way he could keep it together and it all scattered from the table. In some places mirrors felt from the walls...*

2. Gazeta Wileńska, 1786, Dec. 23, No. 51
The same repeated after Gazeta Warszawska, 1786, Dec. 13, No. 9

3. Schlesische Privilegirte Zeitung, 1786, Dec. 20, No. 150
...wall sconces and mirrors fell down in the castle...

4. Lwowskie Tygodniowe Wiadomości, 1786, Dec. 23, No. 95
Wielkanoc is only mentioned that there the earthquake was heard and felt very much.

Wiślica, Poland

1. Kraków AKMK-5, manuscript
... An earthquake extended beyond the whole county of ***Wiślica****... at the same time it was felt more strongly in some places and less in others...*

Włoszczowa, Poland

1. Gazeta Warszawska, 1786, Dec. 13, No. 99
...In ***Włoszczowa*** *(about 12 miles from Kraków) not only at the manor house and in the town, but also in the neigbourhood, chairs moved out from under seated, glasses broke, doors opened on their own, clock bells rang extraordinarily, scarves hung out fell from the ropes, fowl in the fields flew out from under eaves, the domestic fowl dropped off roosts. In the* ***Włoszczowa*** *church and in the manor house cornices fell off...* (1 Polish mile = 7146 m)

2. Gazeta Wileńska, 1786, Dec. 23, No. 51
The same repeated after Gazeta Warszawska, 1786, Dec. 13, No. 99

3. Lwowskie Tygodniowe Wiadomości, 1786, Dec. 23, No. 95
Włoszczowa is only mentioned; there, the earthquake could be heard and felt strongly.

4. Schlesische Privilegirte Zeitung, 1786, Dec. 20, No. 150
after Gazeta Warszawska, but abbreviated

Wolbrom, Poland

1. Kraków AKMK-8, manuscript
.. *From all the parishes of the* [Wolbrom] *Deanery I received the news about the earthquake ... The whole of my parish experienced the effect of this terrible earthquake, to a greater or lesser extent depending whether the places were located higher or lower...*

Wołczyn (Constadt), Poland

1. Schlesische Provinzialblätter, 1786, p. 595
Wołczyn is only mentioned; tremors were felt there but less strongly than in Brzeg, Nysa, Koźle, Racibórz, Prudnik, Krapkowice, Głogówek, Opole, Lewin Brzeski, Stolec and Ziębice.

Wrocimowice (Rocimowice, Racimowice), Poland

1. Gazeta Warszawska, 1786, Dec. 16, No. 100
...In many places near Kraków, cornices and crosses fell from churches - it happened in the parish church at Wrocimowice...

2. Gazeta Warszawska, 1786, Dec. 23, No. 102
*...At **Wrocimowice**, during this earthquake, a metal sheet that covered the dome of the church fell down.*

3. Lwowskie Tygodniowe Wiadomości, 1787, Jan. 03, No. 98
mentioned after Gazeta Warszawska, 1786, Dec. 23, No. 102

Wrocław (Breslau), Poland

1. Schlesische Privilegirte Zeitung, 1786, Dec. 09, No. 145
... An earthquake was noticed by several people on Sunday around 5 p.m, especially in the Oława suburb, Piasek quarter and the cathedral. Some reliable persons assure that they saw suspended clocks and glasses move...

2. Schlesische Provinzialblätter, 1786, p. 595
*... this earthquake was felt in **Wrocław** by various persons, in particular those who were alone at the cathedral, on the bridge, in the Oława suburb, in Albrecht Street (Wita Stwosza Street) and Naschmarkte (Nanker Square). Most people only noticed the slight shaking of doors and that cups and glasses rattled...*

3. Gazeta Warszawska, 1786, Dec. 30, No. 104
   after Schlesische Privilegirte Zeitung, 1786, Dec. 09, No. 145

4. Hoff, 1841, p. 78
   mentioned

5. Jeitteles, 1860, p. 330
mention after Schlesische Privilegirte Zeitung, 1786, Dec. 09, No. 145

Wrzępia, Poland

1. Uście Solne - manuscript
The earthquake was felt in Wrzępia as much as in Uście Solne. For detailed description of the event see Uście Solne.

2. Bójko, 1911, p. 49
after Uście Solne - manuscript

Zator, Poland

1. Gazeta Warszawska, 1786, Dec. 16, No. 100
... There was also an earthquake in Busko, Kielce, Chęciny, Głębowice and the whole of Oświęcim and ***Zator*** *Duchies and closer to Kraków than stronger ... The news came from Zator that the local doctor, J.P.Gherri, an Italian, born in Bologna, knowing from the experience in his native country, that there such events are often not singular but may be repeated, did not sleep the whole night and testified, that there was another earthquake at 3 a.m.,but a weak one which those asleep did not notice. ...*

2. Gazeta Wileńska, 1786, Dec. 23, No. 51
The same repeated after Gazeta Warszawska, 1786, Dec. 16, No. 100

3. Carosi, 1789, p. 37
...The same underground noise, which preceded the earthquake, and the next shakes were felt in Tarnów, Myślenice, ***Zator****, Oświęcim, Bielsko-Biała, and many other places...*

Zduny, Poland

1. Schlesische Provinzialblätter, 1786, p. 595
...It was stronger in Bierutów (than in Oleśnica), however it was far weaker in area of Dobroszyce, Trzebnica, Twardogóra, Milicz, Sułów, ***Zduny,*** *and Rawicz. In most of these localities - apart from the fear of those who experienced it - it was manifested by the slight shaking of chairs, tables, windows in rooms, and all standing objects. Also at this occasion, it was noticed that birds and other animals showed the same signs of fear as those in February...*

Ziębice (Münsterberg), Poland

1. Schlesische Provinzialblätter, 1786, p. 595
Ziębice is only mentioned that there the earthquake was strong but not as much as at Bytom, Tarnowskie Góry, Świerklaniec and Pszczyna.

Žilina, Slovakia

Martinka, 1927, p. 152
December 3, 1786 – theret was an earthquake at ***Žilina,*** *Košice, Spišská Nová Ves, Spišske Podhradie with 3 tremors that caused chimneys fell down*

Żębocin (Zemboczyn), Poland

1. Gazeta Warszawska, 1786, Dec. 23, No. 102
*...At **Zemboczyn** a quater of a mile from Kowala, the local old church, but one which was well kept and decorated, was greatly cracked both in walls and ceiling...*

2. Lwowskie Tygodniowe Wiadomości, 1787, Jan. 03, No. 98
mention after Gazeta Warszawska, 1786, Dec. 23, No. 102

Acknowledgements Barbara Guterch is very grateful to Dr. Morag M. Macdonald, her Scotish friend, for help in translating the ancient Polish documents into English.

References

Bielsko-Biała AP, manuscript, Archiwum Państwowe w Bielsku, Gerichts Protocoll 1785-87, fasc. 326, fol. 129 (in German)
Bójko J (1911) Okruszyny z Gremboszowa, Lwów, p.49 (in Polish)
Brünner Zeitung (1786) No 99, 787; No 101, 799–800; No 102, 811; No 103, 817–818; No 104, 825–826., (Brno, Statni Oblastni Archiv, Beilage zu Brünner Zeitung, cod.I440), Brno (in German)
Carosi JF (1789) Extrait d'une lettre de M. de Carosi, Capitaine au service du Roi & de la république de Pologne, & Correspondant de l'Académie, á Mogila près de Cracovie, in: Nova Acta Academiae Scientiarum Imperialis Petropolitanae, Petersburg, vol. V, 35–38 (in French)
Chrościński I (1896) Izydora Chrościńskiego Kronika z lat 1784–1794 z przydaniem aktów fundacyjnych kolonii akademickiej w Tarnowie z lat 1559 i 1578, jako też Raportu o jej stanie z roku 1766, editor: J. Leniek, Kraków, p. 79 (in Latin)
Cieszyn AP, manuscript, Achiwum Państwowe w Cieszynie (State Archives in Cieszyn), V Abtheilung des Gedenk-Buchs der Stadt Teschen vom Jahre 1767 bis 1822, fasc. 74, fol. 114 (in German)
Droździk W (1979) Z dziejów Kęt, Kraków, p. 84 /in Polish/
Gazeta Warszawska (1786) Dec. 09, No 98-supplement; Dec. 13, No 99; Dec. 16, No 100; Dec. 23, No 102; Dec. 30, No 104, Warszawa (in Polish)
Gazeta Wileńska (1786) Dec. 16, No 50-supplement; Dec. 23, No 51;, Wilno /in Polish/
Grabowski A (1900) Kraków i jego okolice / opisał historycznie Ambroży Grabowski; Friedlein DE (ed), Kraków, pp.187–188 (in Polish)
Grabowski A (1909) Wspomnienia; Estreicher S (ed), Kraków, T. I (In Polish)
Gwiazdka Cieszyńska (1929) April 12, No 28, 4–5, Cieszyn (in Polish)
Hoff KEA (1841) Chronik der Erdbeben und Vulkanausbrüche, Gotha, 78 (in German)
Jeitteles HL (1860) Versuch einer Geschichte der Erdbeben in den Karpathen und Sudeten Landern bis zu Ende des achtzenhten Jahrhunderts, Zeitschrift d. Deutsch Geol Ges 12:287–349 (in German)
Kaleta R (1964) O trzęsieniu Ziemi w Krakowie w 1786 roku. Czasopismo Geograficzne 35(1):96–101, Wrocław (in Polish)
Kárník V, Michal E, Molnàr A (1958) Erdbebenkatalog der Tschechoslowakei bis zum Jahre 1956. Geofysikalni Sbornik 1957, Travaux de l'Institut Geophysique de l'Academie Tchecoslovaque des Sciences, No. 69, pp 444–445 (in German)
Kęty MO manuscript, Muzeum Okręgowe, Oddział w Kętach, Księga Cechu Tkackiego Barchanickiego, fasc. H/697 (in Polish)

Koláček F (1921) Karpatská zemětřesení na pudě Republiky Československé, in: Hostinský B (ed) Publications de la Facuté des Sciences de l'Université Masaryk, No. 6, Brno (in Czech)
Kraków AKAP-1, manuscript, Archiwum Prowincji Zakonu O.O. Kapucynów w Krakowie, Chronologia seu Historia Conventus Rosvadoviensis, fasc. AKR. 26, T. I, fol. 97 (in Latin)
Kraków AKAP-2, manuscript, Archiwum Prowincji Zakonu O.O. Kapucynów w Krakowie, Kronika Klasztoru Krakowskiego, fasc. AKK. 95, fol. 240 (in Latin)
Kraków AKK, manuscript, Archiwum Kapituły Katedralnej w Krakowie, Acta Actorum Capituli 1776–1787, fasc. 25, fol. 714–715 (in Latin)
Kraków AKMK-1, manuscript, Archiwum Kurii Metropolitalnej w Krakowie, fasc. A. Ep. 112, List Pasterski Prymasa Polski, fol. 203 (in Polish)
Kraków AKMK-2, manuscript, Archiwum Kurii Metropolitalnej w Krakowie, fasc. A. Ep. 112, Zalecenie do kościołów farnych i zakonnych, fol. 252–253 (in Polish)
Kraków AKMK-3, manuscript, Archiwum Kurii Metropolitalnej w Krakowie, fasc. Ag. 10, Acta Gratiosa, fol. 72 (in Polish)
Kraków AKMK-4, manuscript, Archiwum Kurii Metropolitalnej w Krakowie, Raporta ex Cancellaria Oficij Consistorij Sandomiriensi ex Mense Decembri Anni 1786, fasc. Ag. 12, fol. 386 (in Latin)
Kraków AKMK-5, manuscript, Archiwum Kurii Metropolitalnej w Krakowie, fasc. Ag. 12, (Raport z Oficialiatu Wiślica), fol. 388 (in Latin)
Kraków AKMK-6, manuscript, Archiwum Kurii Metropolitalnej w Krakowie, fasc. Ag. 12, Raport Miesiaca Grudnia z Dekanatu Kijskiego Roku 1786, fol. 391 (in Polish)
Kraków AKMK-7, manuscript, Archiwum Kurii Metropolitalnej w Krakowie, fasc. Ag. 12, Raport z Dekanatu Koprzywnickiego w Miesiącu Grudniu RP. 1786, fol. 393 (in Polish)
Kraków AKMK-8, manuscript, Archiwum Kurii Metropolitalnej w Krakowie, fasc. Ag. 12, Raport Dekanatu Wolbromskiego dnia 31. Grudnia 1786, fol. 395 (in Polish)
Kraków AKMK-9, manuscript, Archiwum Kurii Metropolitalnej w Krakowie, fasc. Ag. 12, (Raport z Parafii Kromołów), fol. 397 (in Polish)
Kraków AKMK-10, manuscript, Archiwum Kurii Metropolitalnej w Krakowie, fasc. Ag. 12, Rapport z Parafyi Kościoła Mrzygłodzkiego Mil dziesięć od Krakowa to Miasto położone, fol. 399 (in Polish)
Kraków AKMK-11, manuscript, Archiwum Kurii Metropolitalnej w Krakowie, fasc. Ag. 12, Z Parafii Potockiey Dekanatu Lelowskiego, fol. 401 (in Polish)
Kraków AKMK-12, manuscript, Archiwum Kurii Metropolitalnej w Krakowie, fasc. Ag. 12, Rapport Dekanatu Nowogórskiego w Miesiącu Styczniu Roku 1787, fol. 465 (in Polish)
Kraków AKMK-13, manuscript, Archiwum Kurii Metropolitalnej w Krakowie, fasc. Ag. 12, (Report of Trzebinia parish), fol. 469 (in Polish)
Kraków AP-1, manuscript, Archiwum Państwowe w Krakowie (State Archives in Kraków), Archiwum O.O. Augustianów, fasc. 22, Kronika Klasztoru, fol. 88, fol. 109 (in Latin)
Kraków AP-3, manuscript, Archiwum Państwowe w Krakowie (State Archives in Kraków), Teki Schneidra, fasc. 1623, (earthquake in 1786 – reports of local authorities in Myślenice, Nowy Sącz, and Tarnów) (in German)
Kaków AP-4 manuscript, Archiwum Państwowe w Krakowie, Księga radziecka, fasc. K -329, fol. 541 (in Polish)
Kraków AREF, manuscript, Archiwum Prowincji Zakonu O.O. Reformatów w Krakowie, Kronika Klasztoru Krakowskiego, fasc. II, fol. 203 (in Latin)
Kraków-MOGIŁA AOC, manuscript, Archiwum Opactwa O.O. Cystersów, Notatnik zdarzeń kleryka mogilskiego 1785–1789, fasc. 11, fol. 47 (in Latin)
Kreutz F (1876) Rzecz o trzęsieniu ziemi w Galicji wschodniej z 1875 r., Kosmos, Lwów 1 (in Polish)

Krzyżanowski T (1900) Wspomnienia mieszczanina krakowskiego z lat 1768–1807; Prokesch W (ed), Kraków, 29–30 (in Polish)
Laska W (1902) Die Erdbeben Polens, Mittheilungen der Erdbeben Commission, Keis. Ak. d. Wiss. in Wien, Wien, No.8, 1–36 (in German)
Lwowskie Tygodniowe Wiadomości, 1786, Dec. 20, No 94; Dec.23, No 95; Dec.27, No 96; Lwów (in Polish)
Lwowskie Tygodniowe Wiadomości, 1787, Jan. 03, No 98, Lwów (in Polish)
Mallet R, Mallet JW (1858) The earthquake catalogue. London, p 15
Martinka J (1927) Seizmologicky katalog Slovenska, Sbornik muzealnej spolecnosti slovenskej, c. 3–4, katalog Slovenska, p 144–156 (in Slovakian)
Olczak T (1966a) Sur la region epicentrale et la profondeur d'un foyer seismique du 3 XII 1786 dans les Carpathes occidentales polonaises. Acta Geophys Pol 14(1):61–67 (in French)
Olczak T (1966b) O Karpackim trzęsieniu ziemi z dnia 27 II 1786 roku i towarzyszących mu ruchach sejsmicznych na przedpolu Sudetów Wschodnich, Przegląd Geologiczny, XI (XIX), No. 1, 3–21 (in Polish)
Pagaczewski J (1964) Trzy silne trzęsienia ziemi w Krakowie i okolicy w latach 1785–1786, Wszechświat, No 2, 39–42 (in Polish)
Pagaczewski J (1972) Catalogue of earthquakes in Poland in 1000–1970 years. Publs Inst Geophys Pol Acad Sci 51:3–36
Pamiętnik Historyczno Polityczny (1787) Pamiętnik historyczno-polityczny przypadków, ustaw, osób, miejsc i pism wiek nasz szczególniey interesujących, 1786; Świtkowski P (ed), Warszawa, vol. IV, 237–238, Warszawa (in Polish)
Perrey A (1846) Memoire sur les tremblements de terre dans le bassin du Danube, Lyon, p 361 (in French)
Remeš M (1902) Zemětřesení na Moravě pozorovaná. Věstník klubu přírodovědeckého v Prostějově, roč. V., 8
Réthly A (1952) A Kárpátmedencsék Földrengésci 455-1918, Akademiai Kiado, Budapest (in Hungarian)
Schlesische Privilegirte Zeitung (1786) Dec. 09, No 145; Dec. 11, No 146; Dec. 20, No 150; Dec.23, No 151, Wrocław (in German)
Schlesische Provinzialblätter (1786) vol. 4, No 12, 593–596, Oleśnica (in German)
Sieberg (1940) Beitrage zum Erdbebenkatalog Deutschlands und angrenzender Gebiete für die Jahre 58 bis 1799. Mitteilungen des Deutschen Reichs-Erdbebendienstes, Jena, No. 2 (in German)
Tarnów AD, manuscript, Archiwum Diecezjalne w Tarnowie, Notificationes Anno 1781–1798 (in Polish)
Tarnów MO, manuscript, Muzeum Okręgowe w Tarnowie, Acta Advocatalia et Scabinalia Tarnoviensia, fasc. MT-H 496, fol. 44 (in Polish)
Trzciński A (1787) Opisanie przyczyn fizycznych Trzęsienia Ziemi, Kraków, 24–27 (in Polish)
Uście solne, manuscript, Kościół Parafialny w Uściu Solnem, Liber Copulatorum et Liber Baptisatorum (since year 1775), fol.49 (in Polish)
Wojciechowski B (1895) Fragment z dziejów szkół kaliskich, in: Archiwum do dziejów literatury i oświaty w Polsce, Kraków, vol VIII, 323–324 (in Polish)
Wrocław BOS, manuscript, Biblioteka im. Ossolińskich, fasc. 11634. II, (Raport pułkownika Dawida Grammlicha do generala Jozefa Wodzickiego), fol.985-986 (in Polish)

CPSIA information can be obtained at www.ICGtesting.com
Printed in the USA
LVOW01*2128120515

438213LV00001B/36/P

9 783319 154459